J. Batty Tuke

The Insanity of over-exertion of the Brain

Being the Morison Lectures

J. Batty Tuke

The Insanity of over-exertion of the Brain
Being the Morison Lectures

ISBN/EAN: 9783337075101

Printed in Europe, USA, Canada, Australia, Japan

Cover: Foto ©berggeist007 / pixelio.de

More available books at **www.hansebooks.com**

THE INSANITY

OF

OVER-EXERTION OF THE BRAIN.

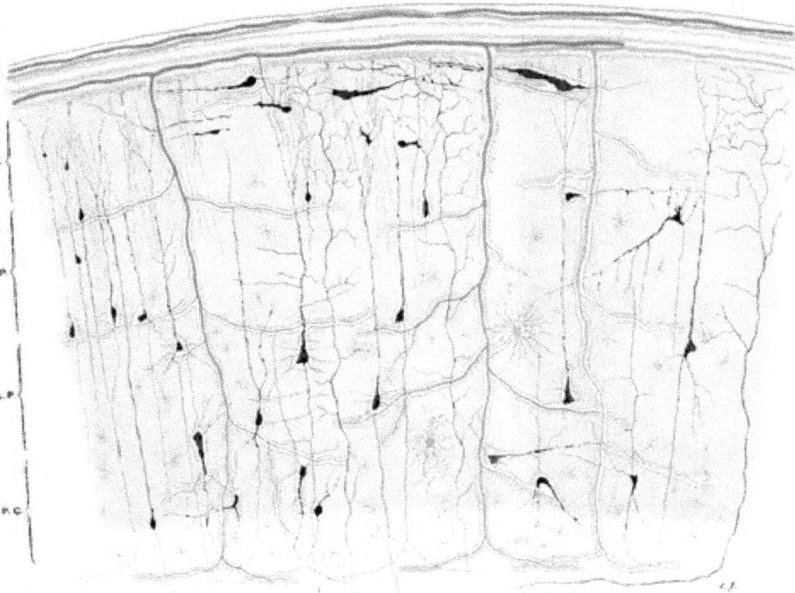

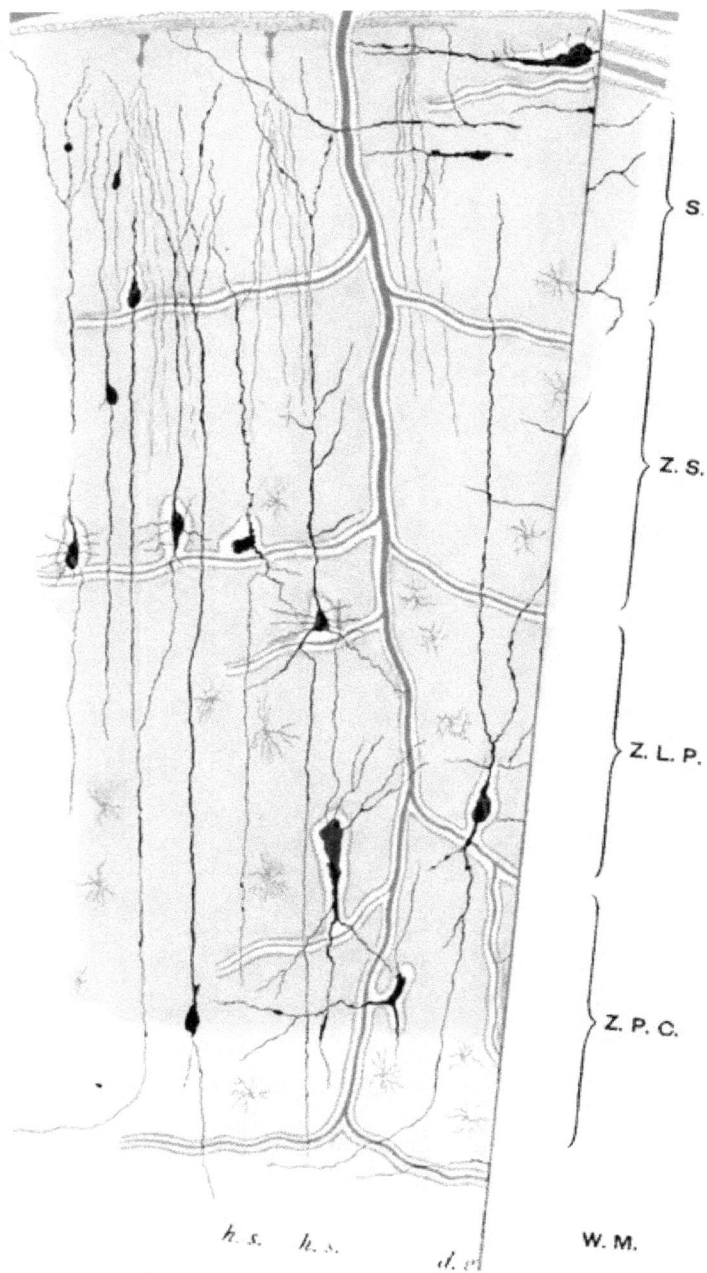

S.

Z. S.

Z. L. P.

Z. P. C.

h. s.   h. s.   d. e.   W. M.

THE INSANITY

OF

Over-Exertion of the Brain;

BEING

THE MORISON LECTURES

DELIVERED BEFORE THE

ROYAL COLLEGE OF PHYSICIANS OF EDINBURGH,

Session 1894.

BY

J. BATTY TUKE, M.D., F.R.C.P.E., F.R.C.S.E.

WITH ILLUSTRATIONS AND DIAGRAMS.

EDINBURGH:
OLIVER AND BOYD, TWEEDDALE COURT.
LONDON:
SIMPKIN, MARSHALL, HAMILTON, KENT, AND CO., LIMITED.

PREFATORY NOTE

SINCE the publication of the following Lectures in the *Edinburgh Medical Journal*, important investigations have been made, bearing on certain opinions and theories advanced in them. As the results of these investigations have been given to the public only in the form of Lectures or Demonstrations, I deem it advisable not to anticipate their fuller publication. From what has been communicated to me, I feel, however, justified in saying that these researches will throw much light on the nature of cell changes, on the rapidity of subsequent deterioration of the connection-apparatus, and on the absolute necessity of rest for recuperation ; and will thus aid us to arrive at a rational system of treatment of the Insanities.

Much honour must accrue to HODGE of the Clark University for initiating a system of experimentation which opens up a novel and important line of enquiry. To have demonstrated that brain cells are, even as other cells are, liable to diurnal physiological waste and restoration, is a matter of deep interest to the physiologist and psychologist. My hearty thanks are due to him for the generous manner in which he placed his preparations at my disposal.

I beg also to acknowledge the great assistance afforded me by LOVELL GULLAND, GUSTAV MANN, LLOYD ANDRIEZEN, and MILES, who, in the truest spirit of scientific unselfishness, presented me with specimens for the purpose of illustration.

In the recently published *Nouvelles Idées sur la Structure du Système nerveux chez l'Homme et chez les Vertébrés*, Ramon y Cajal has modified and amplified certain of the descriptions given in the *Nuevo Concepto de la Histologia de los Centros Nerviosos.* This applies specially to the description of the external and internal cortical zones. The reader is referred to the former work for full details.

SAUGHTON HALL.
    GORGIE, EDINBURGH.
        July 30. 1894

THE INSANITY

of

OVER-EXERTION OF THE BRAIN.

LECTURE I.

Mr President and Gentlemen,—My intention is to study in the following course of lectures the methods of action of certain of the proximate causes of insanity, and to indicate what I believe to be the principles of treatment founded on these considerations. It will be manifestly impossible to cover the whole ground, and I shall therefore restrict myself mainly to investigating the *modus operandi* of one of the most important morbid influences productive of insanity,—over-exertion of the brain ; and, further, shall confine myself almost exclusively to the study of its action during the earliest stages of the condition produced ; the stages during which the practitioner of medicine has for the most part to deal with them, and during which the deranged cerebral apparatus is in a state to react to therapeutic measures.

I find my general position in regard to the insanities so well stated by Creighton in the opening sentences of his article on " Pathology " in the ninth edition of the *Encyclopædia Britannica,* that it would be sheer piracy were I to attempt to render his expressions in my own words. I therefore quote them, —" Although it is laid down, in the opening sentences of the Hippocratic treatise *de prisca medicina*, that the medical art, on which all men are dependent, should not be made subject to the influence of any hypothesis, that the care and cure of the sick should not be subordinated to patho-

A

logical theory, but should be guided by experience; yet
the practitioners of medicine have at no time been able to
dispense with theory, not even those avowed followers of the
Hippocratic tradition who, while they professed a kind of
quietism amidst the rise and fall of systems, have none the less
been profoundly influenced by theory at every step of their
practice. The position of Cullen is the only rational one: 'You
will not find it possible to separate practice from theory altogether;
and, therefore, if you have a mind to begin with theory, I have
no objection. . . . . *To render it safe, it is necessary to cul-
tivate theory to its full extent.*' The progress of pathology hitherto
has been exactly parallel with the progress of philosophy itself,
system succeeding system in genetic order. No other department of biological science has shown itself so little able to
shake off the philosophical character, or to run in the career of
positivism or pure phenomenalism. This unique position of
pathology among the natural sciences is doubtless owing to the
fact that it is a theory of practice,—a body of truth and guess-
work existing for the benefit of a working profession which is
daily brought face to face with emergencies and is constantly
reminded of the need of a reasoned rule of conduct."

It is on these lines that I propose to work, and in so doing it
will be necessary to touch here and there on certain points which
were partially dealt with in the course of Morison Lectures
delivered by me in this hall some twenty years ago, and to
endeavour to bring into line the results of certain observations
and theories which have been made and advanced in the interval.

In a popular article published some years ago in the *Nine-
teenth Century*, I endeavoured to show the social influences
which had induced the public and the profession to regard
insanity as a disease of the mind. The prominence of
the symptoms, the almost total ignorance of brain anatomy,
physiology, and pathology, and the necessity of protection of
the public and the lunatic, combined to direct the feelings of all
concerned in the direction of insanity being regarded primarily
as a perversion of the intellect. The popular conception of the
condition had been from the earliest ages down to a not very
remote period entirely psychological; at the best it had been
but a rude mixture of a pseudo-psychology and of a pseudo-

pathology, identical with what we find in the pages of Burton. Medical science had not been in a position to exercise any definite influence on popular views from a strictly professional standpoint, inasmuch as the physician's conception of the condition had been in no way different in degree or in kind from that of the general public. The views of Medicine regarding insanity at the end of last century remained practically unchanged from those of the Middle Ages; its character as a symptom of disease being pretty generally recognised, but the recognition being accompanied by the feeling that it differed in some mysterious way from all other diseases. It was therefore inevitable that the psychological principle should find expression in nomenclature, and that the original Hippocratic classification of mania, melancholia, and dementia should be adhered to down to within the last thirty years. Let me cite (or, rather, re-cite) a prominent example of the long and strong survival of the psychological principle. In the first issue (1871) of the official " Nomenclature of Diseases drawn up by a Joint Committee appointed by the Royal College of Physicians of London," all the insanities were ranged under six " Disorders of the Intellect," a severe line being drawn between them and diseases of the general system. Therapeutic treatment was necessarily affected by this crystallization of false theory, and was conducted mainly on humanitarian lines. In point of fact the spirit of scientific medicine lay almost dormant in so-called alienistic practice.

Concurrently (or, at least, almost concurrently) with the advances made in nervous anatomy, physiology, and pathology during the last thirty-five years, the views of the specialty of so-called psychiatric medicine have become to some extent altered or modified as regards nomenclature. Etiology has, in theory, if not in practice, been adopted instead of mere symptomatology as a basis of classification; and, of late years, the habit has been growing of connecting in nomenclature what is believed to be the cause of the morbid mental symptoms with the general character of the symptoms themselves. The various etiological nomenclatures which have been advanced have served the good purpose of keeping before the mind that insanity is merely a symptom of somatic disease (using the word disease in its widest sense); by deflecting the

observer from the misconception involved in the term " Mental
Disease," and by focussing his attention on the phenomena
following on actual or presumed causes, they have directed
inquiry as to consequent alterations in tissue. They have acted
as guides to the higher pathological platform, to the study of
the changes in the apparatus of the central organs through which
consciousness is manifested, and the implication of which we
have every right to assume is productive of morbid mental
phenomena. There is every probability that we will have to
retain this system of classification for many years to come as a
provisional arrangement, notwithstanding that grave errors are
apt to arise from its being employed on merely *prima facie*
evidence of the connection of cause and effect. There is even a
possibility that its terminology may become permanent in consequence of a definite pathological meaning becoming attached
to many of the presently recognised etiological groups. It is
only too true that we have not as yet advanced very far on the
road ; but, when such work as that of Lockhart Clark, Betz,
Deiters, Meynert, Golgi, Ramon y Cajal, Obersteiner, Retzius,
Bevan Lewis, and others, is considered, it is evident that we have
some reason for the hope that the study of the insanities will in
future be conducted on the same scientific principles that govern
our views of disease at large.

I have sometimes speculated, and it seems to me not unprofit-
ably, upon the attitude which the profession would assume if a
case of so-called acute mania, or of so-called acute melancholia,
were introduced to it for the first time. In order to place you
in the same position for speculation, I am going to ask you to
try to put out of your minds that you have ever heard or
read of such cases, to make a *tabula rasa* of experience, opinion,
and feeling, in respect of such insanities. Let us try to imagine
that two cases are introduced to us presenting a congeries of
symptoms which had not previously come under our observation,
the most prominent in each being a degree of mental alienation.
The first, a man whose general bodily condition is poor, muscles
flabby, face pale and wearing an anxious expression, pulse quick
and irritable, temperature slightly above the normal, skin clammy,
blood reduced as regards number of corpuscles and amount of

hæmoglobin, bowels costive, oxalates or phosphates in large quantities in urine ; he is reported to be sleepless, the whole system is generally out of tone, but no active disease of any organ is detectable : he is wringing his hands, is restless, cries out that his soul is lost, and, in a word, is deeply melancholic, and is reported to have attempted suicide. The second case—a woman—presents much the same bodily symptoms, but in addition the temperature is higher, and menstruation has been arrested ; mentally she is exalted, as shown by general and verbal expression, and she may be the subject of some extravagant delusion. It is stated that in both cases the patient had suffered for long from serious anxieties and sorrows.

I submit that, under the circumstances we are supposing, the mind of the College would not turn to the idea contained in the term "mental disease," and that first impressions would lead to the rough provisional diagnosis of obscure bodily disease, accompanied by unusual forms of delirium. Our clinical experience would have led us to refer certain secondary affections of the brain to certain definite morbid conditions of the general system or of special organs. Our minds would have reverted to the euphoria of phthisis, the depression of gout, the delirium of fever, the coma of Bright's disease, and the excitement following on the administration of certain poisons. It appears to me, therefore, extremely likely that the mental symptoms would not bulk very largely in our eye, that we would commence the study of this new congeries of symptoms by careful investigation into the bodily condition of the patients, and would not begin by attempting to analyze the mental symptoms : we would regard them as mere incidents in the case, exactly as we had been in the habit of regarding muttering, noisy, or joyous delirium, which we had observed as a secondary consequence of various diseases. After careful examination of all the systems, we would be forced, by a process of exclusion, to adopt the rough provisional diagnosis of disease of the brain accompanied by mental symptoms,—a very different diagnosis, I submit, from that of mental disease. In what part of the organ would we localize disease? It is needless here to narrate our reasons, founded on physiology, for believing that the frontal and superior convolutions form the substrata through which psychical

action is manifested ; or those based on pathology for our belief
that implications of these gyri are followed by abnormal exercise
of function.   Suffice it to say, that all our teaching would lead
us to localize the presumed disease in that region.   Turning to
the history of the cases, the only assignable reason for the
existence of bodily and mental symptoms is found to be prolonged painful emotion.   This, although it might suggest itself
as a cause of melancholy, could not, to men brought up in the
atmosphere of physiology and pathology, afford any sufficient
reason for the systemic degradations.   We would, therefore, I
submit, push our inquiries as to the manner in which the cause,
emotion, could so act on the frontal and superior gyri of the
brain as to produce morbid modifications of the general
trophesis, and of the mental functions in the direction of exaltation and depression.

At this point I will abandon the parable for a time, in the
hope that I shall be able to resume it after having laid before
you certain facts and theories bearing on the general proposition
that insanity is the result of well-defined morbid processes, and
that it is but one of a group of symptoms in a given case, the
relative value of which, from a pathological and therapeutic
point of view, has to be carefully weighed.

Although we have too often to acknowledge that our knowledge of the Institutes of Medicine is far from perfect, and have
to recognise it in terms by speaking of " functional " diseases in
connection with morbid symptoms which we have not as yet
been able to refer definitely to actual structural changes, still
science is gradually narrowing the area of ignorance, and we are
able year by year to base diagnosis and treatment more fully
and more firmly on a definite knowledge of the normal and
morbid anatomy and physiology of involved organs.   This
platitude is specially applicable to the heart, lungs, and kidneys,
the treatment of the diseases of which organs has very materially
advanced in consequence of direct scientific investigation.   It is
not much more than thirty years since similar methods of
research were applied to the brain, and already their effects in
the treatment of many cerebral diseases have been practically
demonstrated, and are so well known to all present that they
need not be further alluded to.   But I ask you whether,

as regards the insanities, theory founded on this new knowledge has been pushed far enough to seriously influence
treatment, and whether the relations one to the other of the
various organs which compose the substance of the brain
have been seriously studied with the view of considering in
what manner, and to what extent, implications of one or more
of them may react on the functional activity of the organ. I
take the liberty of believing that the practitioner of medicine
is not so intimately acquainted with the minute anatomy of a
convolution as he is with that of the kidney or liver, and that
he is therefore not able to visualise the probable changes which
may occur in a gyrus during the action of disease. I believe
that every thinking man, in the presence of disease, projects a
scheme of the departures from healthy structure on a mental
screen—whether rightly or wrongly does not affect my position ;
he visualises (as Galton calls it) the altered condition of matters ;
he sees the gross lesions and the microscopic details of the
changes in the various structures of the organ, and the picture
influences him in the future conduct of the case. This simply
means that he brings his anatomical, physiological, and pathological learning to bear on an individual instance, and in so
doing he forms a conception of the nature of the implication of
the apparatus. But I again take the liberty of asking—Is a brain
convolution often regarded as an apparatus, and is the same
kind of conception arrived at in respect of the possible and
probable change which may take place in it in consequence of
the action of morbific influences ? Is the normal structure before
the mind's eye as it is in the case of, let us say, the kidney?
Believing that it is not so, I shall attempt to diagrammatise a
convolution as an organ. To those who are acquainted with the
subject I must apologise ; but as I long experienced difficulty
in realizing to myself the relations of the various structures one
to the other, I venture to think that the same may have been
felt by others.

The main obstacle to forming a mental diagram of the
structure of a convolution is that, although text-books give, with
one important exception, accurate descriptions of the individual
elements, they fail to bring them together so as to enable the
student to appreciate the lymphatics, the bloodvessels, the fibres,

the endothelium, and the connective tissues, and accordingly he is not able to grasp at once the idea of the possible results of morbid changes occurring in one or more of these structures.

In speaking of the arrangement of the cortical cells I follow the description of Ramon y Cajal, which, being apparently accepted by Retzius, may be considered to be the latest authoritative statement. So far as the cells are concerned, the diagram you hold in your hands is founded on the figures given by these two investigators. In describing a typical Rolandic convolution, Cajal speaks of four layers of cells, the two inner layers, which have been generally accepted as separate, being merged in one. The letters S. Z., Z. S. P., Z. L. P., and Z. P. C. apply to the superficial zone, the zone of small pyramids, the zone of large pyramids, and the zone of polymorphous cells. Till lately the superficial layer was believed to contain only a few small fusiform cells surrounded by numerous nuclei of neuroglia. But Cajal has demonstrated, by a modification of Golgi's process, the presence of other cells peculiar in character. He describes as existing in this zone cells of four types,—polygonal, fusiform, triangular, and unipolar. The polygonal cells are the least numerous, and their poles branch in various directions ; the fusiform, also not numerous, the long axis of which lies parallel to the plane of the surface, give off an apical and a basal process, which does not arise directly from the body of the cell, but from the protoplasmic expansions ; these poles run horizontally or slightly diagonally for great distances, never descend into the inferior layers, and give off collaterals (certain of which tend towards the surface) which form an exceedingly rich plexus of terminals ; in some cases the axis-cylinder process breaks up into two or three branches, and gives off collaterals ; the more numerous pyramidal cells give off two or more axis-cylinder processes which mostly run to the surface.

The second zone, that of small pyramids, sends its apical processes to terminate in "splendid tufts" in the meshwork of the superficial layer, thus constituting a very thick protoplasmic plexus. Before doing so it sends off numerous collaterals, as is also the case with the axis-cylinder process, which, as in the case of the cells of the outer layer, divides into two twigs. The lateral expansions are numerous, extend for long distances, and end without anastomosing in similar expansions of other cells. The plexus so formed corresponds to the broad stripe of Baillarger.

The cells of the layer of large pyramids resemble the small pyramids, only differing in size and in the greater thickness of their apical and axis-cylinder poles. These "giant cells" of Betz occur principally near the vertex; as we examine forwards or backwards they fall off in size and number, and are not seen in the occipital lobe. The plexus formed by the collaterals of their axis-cylinder processes corresponds to the thin stripe of Baillarger.

The chief characteristic of the layer of polygonal cells (which are described as egg-shaped, spindle-shaped, triangular or polygonal) is that the apical process never reaches the superficial zone. The axis-cylinder process in descending gives off three or four collaterals, which again ramify, and unites itself, either by bending or by a ⊥-shaped division, with one or two nerve-fibres of the white substance.

Golgi describes, further, as mingled in small numbers in the second, third, and fourth layers, but especially in the fourth, certain "cells of short axis-cylinder" which terminate in ramifications through the thickness of the grey substance, and (according to Cajal) enclose the bodies of neighbouring cells. In the same situation, and also sparsely scattered, Marinotti describes spindle-shaped cells which send processes mostly to the superficial layer, and (according to Cajal) occasionally to the zone of small pyramids. Golgi figures solitary cells occurring in all the layers, but especially in the fourth, the protoplasmic processes of which do not reach the surface, and whose axis-cylinder descends for a short distance.

It would be exceedingly interesting to follow out Cajal's views and hypotheses as to the functions of different cells, of the method of their association, and as to the relations of the axis-cylinders of the various layers and those of their collaterals with the projection, association, and commissural systems of fibres in the white matter. At present it is only necessary to say that he holds the projection fibres to be in direct communication with the cells of the pyramidal layers, giving off numerous collaterals running to the corpus striatum, where, by very delicate ramifications, they are connected with its cells. The origin of the

B

callosal fibres is doubtful, according to Cajal ; some may arise
directly from axis-cylinders, others from bifurcating axis-
cylinders of pyramidal cells. In a future lecture we will have
to consider the connection of cell with cell, and convolution with
convolution.

At present it is only necessary to allude to a fourth system
of fibres, which he figures and describes. I quote his own words :
—" In addition to the projection, commissural, and association
systems, there are other and much thicker fibres, proceeding
possibly from the medulla or cerebellum, which, passing usually
in an oblique or horizontal direction through the thickness of
the grey matter, form ramifications of enormous extent throughout its whole substance, the superficial zone included. The
terminal branches form varicose arborisations which appear to
enclose preferentially the small pyramids. Is it possible that
such fibres represent the cerebral termination of sensory nerves,
intermixed with the terminal branches of other nerves ? " This
system he speaks of as centripetal or terminal fibres (c. f. in
Diagram). It will be necessary to revert later on to this subject
and to Cajal's observations and hypotheses ; but, in the meantime, I only desire to draw your attention to the fact of the
immensely intricate meshwork of cell and nerve terminals in
the outer surface of the grey matter of the convolutions.

The body of the cell is encapsulated by a projection of the
hyaline sheath of its nutrient capillary (p. v. s.), the perivascular
space so formed being the commencement of the lymph-path
which leads from the cell to the main lymphatics of the head
and neck. The brain has no true lymphatics, and is dependent
on the perivascular system for the removal of effete and super-
fluous material. As much of what I have to say in future
lectures depends on the thorough understanding of the cerebral
lymphatic system, it will be well to state it definitely here.

If we make a transverse section through the head of a small
animal, sufficiently thin to be observed by a low power, the
arrangement of the membranes is seen to be as follows :—1 (see
Diagram), the dura (D. M.), a thick membrane consisting of two
layers, and containing arteries, veins, and lymphatics ; 2, a fine
membrane, extending over the sulci (A. P.), but intimately connected with, 3, a delicate membrane which invests the grey matter

of the convolutions, and is traceable to the bottom of the sulci
(*V. P.*). Between these two membranes, which are held together
by numerous trabeculæ, is the pial space (*p. s.*), in which blood-
vessels ramify in large numbers. It is, in fact, a highly vascular
double membrane. The chief reason why the lymphatic system
of the brain and spinal cord is not generally understood consists
in the traditional error, to be found in most text-books, of
describing an arachnoid membrane apart from the two layers
just described, and the relations of which are demonstrated in
the specimen submitted to you. In the last edition of Quain's
Anatomy an arachnoid is fully described, and the pia is stated
to be a highly vascular membrane which "dips into all the sulci,
most of which thus contain a double layer."[1] This is directly
contrary to fact : the visceral layer of the pia is in apposition to
the grey matter, and dips into all the sulci ; the parietal layer
bridges the gyri, and between them the vessels are distributed.
No membrane exists between the parietal pia and the dura ;
there is a space lined on both surfaces by endothelium, the sub-
dural space (*s. d. s.*). It would tend to obviate error were the term
"arachnoid" entirely dispensed with, and were all to speak of
the visceral and parietal layers of the pia as we do of those of the
pleura ; but, as a compromise, certain authors employ the term
arachno-pia and true or visceral pia, which terms may be accepted
on the understanding that no membrane exists between the
arachno-pia and the dura. By the arachno-pia bridging the sulci
and the great fissures (such as the sylvian, Rolandic, and intra-
parietal), cisterns are formed between it and the true pia, and at
the base large reservoirs are produced by the arachno-pia bridging the space between the cerebellum and the medulla, between
the pons and the medulla, and between the various eminences
of the cerebrum. The largest, the *cisterna cerebello-medullaris*, is
continuous with the lymph spaces of the spinal cord, which find
vent through all the foramina of the spinal column. In addition,
as shown by Key and Retzius, the dura and the arachno-pia
send prolongations around each vessel or nerve entering or
leaving the skull.[2] All these lymph paths are connected with

---

[1] Quain's *Anatomy*, vol. iii., Part 2, p. 184.

[2] Except the optic and auditory nerves, the lymph currents of which run backwards.

the extra-skeletal lymph system. Another and important method of communication is afforded by the Pacchionian villi, which extend from the arachno-pia through the subdural space, and are in close relationship to the superior longitudinal sinus and the veins of the dura.

Returning to the surface of the convolutions, all its blood-supply is derived from the pial arteries, which on the summits are given off directly, and in the sulci form large descending branches. Each vessel carries with it a prolongation of the pia, which surrounds the vessel to its ultimate branches. This is often spoken of as the adventitia of the vessel. From this pial or hyaline sheath (*h. s.*) processes are given off which encapsulate the cells, each of which is thus surrounded by a perivascular capsule (*p. v. c.*). All these lymphatic paths can be injected from the arachno-pial space in the cord. Mr Miles has kindly supplied me with specimens illustrating this fact; he has succeeded in injecting all the cavities of the brain from the cord, all the cisterns and the perivascular canals of the grey matter of the convolutions; his specimens also show the ease with which the subdural can be injected from the arachno pial space, passing through the Pacchionian villi, which transmit the coloured material to the dura itself. Further evidence of the easy communication between the head and the cord is obtained by opening the spine of a fœtus and pressing and releasing the elastic skull. The cerebro-spinal fluid ebbs and flows with great readiness.[1]

There are reasons for believing that the pericellular capsules are not the actual extremities of the drainage system. Bevan Lewis regards the large connective tissue spider-shaped cells, first described by Deiters, and therefore called after him, as forming the distal extension of the lymphatic system, or, as he otherwise calls it, the lymph-connective system permeating the

---

[1] His describes an epicerebral lymph space as existing between the visceral pia and the surface of the convolution, and extending inwards along the vessels between the hyaline sheath and the brain substance. I have never been able to demonstrate this space in healthy subjects; and it is difficult to understand where the contained lymph could find vent. The only means of reaching the lymphatic system would be by stomata in the visceral pia and hyaline membrane, which have not been demonstrated. The probability is that any such space, when it is found, is artificially produced by the accumulation of morbid fluids. In this space we never find *débris*, as we do in the lumen of the true lymphatic.

neuroglia. Lewis holds that the thick process of these cells terminates in the perivascular or hyaline sheath, and that the finely reticulated processes permeate the neuroglia.

Lewis's opinions have been fully confirmed by Lloyd Andriezen, whose further observations on the character of the neuroglia elements are most important. He divides them into *protoplasmic* and *fibre* elements. The former (*p. g. c.*) are of mesoblastic origin, occur almost entirely in the cortex, possess a distinct cell body, are stellate or dendritic in the arrangement of their processes, which are stout, coarse, and shaggy in outline ; they are connected by one or two of the coarser processes with the hyaline sheath of vessels, and are surrounded by lymph spaces which are in direct communication with the perivascular lymphatic. A smaller variety surrounds the bodies of the greater nerve-cells (pericellular elements). Andriezen believes that they exercise a lymphatic function.[1] The fibre elements are of epiblastic origin, and their function is purely sustentacular. Their body is small, and their processes long, fine, and delicate. They form—(1), a *diffuse network* throughout the cortex and white matter (*d. g. c.*) ; and (2), two *condensation systems,*—the first, on the surface, constituting "a felt-work of tangential fibres interlacing with one another," and sending downwards through half the thickness of the cortex a series of fine fibrils (*s. c. g. c.*) ; the second surround the vessels, especially in the deeper layers of the cortex and the medullary substance ; they are entirely imbedded in the brain substance, and have no anatomical continuity with the hyaline membrane (*v. c. g. c.*). This perivascular condensation is stated to be so great as to practically form a fourth sustaining envelope—a "felted sheath."[2]

This is a rough description of the elements of a convolution, which are fairly represented in the Diagram individually. I now ask you to multiply the items by tens or twenties as regards the vessels, by hundreds as regards the nerve-cells, their processes, and the nuclei of neuroglia ; in the superficial zones, and in those of the small and large pyramids, to extend to the right and left an intricate meshwork of processes ; and throughout all to construct the most delicate fibrillar meshwork proceeding

[1] *British Medical Journal,* 29th July 1893.
[2] *Internationale Monatsschrift f. Anat. u. Phys.,* 1893, Bd. x., Heft 11.

from the nuclei of neuroglia ; to prolong the axis cylinders downwards to the white matter, making some turn inwards to the corpus callosum, others to the right or left to the association system, and others to run directly downwards to the projection expansion. Finally, carry upwards from the *corona radiata* certain thick fibres to terminate on the superficial zone without passing through any cell substance. By so doing you may gain a conception, not merely of the complexity of the organ, but of what is of major importance, a conception of the definiteness of its anatomy, any modification of which must be followed by modification of function.

DIAGRAMMATIC SCHEME OF THE CONSTITUENTS OF A CONVOLUTION.

As regards the cells, fibres, and neuroglia, this diagram is founded on plates by Retzius, Cajal, and Andriezen.

The arrangement of the vessels is purely diagrammatic.

Endothelium is indicated by dotted yellow lines. Lymphatic spaces and channels are left white.

D. M., Dura mater ; *s. d. s.*, Sub-dural space ; *A. P.*, Arachno-pia ; *p. s.*, Pial space ; *V. P.*, Visceral pia ; **S Z.**, Superficial zone ; **Z. S. P.**, Zone of small pyramids ; **Z. L. P.**, Zone of large pyramids ; **Z. P. C.**, Zone of polymorphous cells ; **W. M.**, White matter ; *h. s.*, Hyaline sheath ; *p. g. c.*, Protoplasmic glia cells ; *s. c. g. c.*, Superficial condensation of glia fibre cells ; *v. c. g. c.*, Vascular condensation of glia fibre cells.

LECTURE II.

Mr President and Gentlemen,—In dealing with the anatomy of a convolution I trust it is fully understood that it is not my intention to endeavour to work out the dynamics of a delusion. My purpose is much more humble, being simply to demonstrate how the normal economy of the apparatus may be implicated by the action of disease, and to consider how we may best help to restore healthy relations. As yet the chemistry and mechanics of psychical processes are matters of surmise and theory ; all we can say is, that mental action is a function of connections, or as Obersteiner puts it, "The grey matter is a field for the association of afferent sensory impulses. In it they are placed in communication with efferent paths along which they travel either immediately or at some subsequent time ; or, to speak more correctly, the efferent impulse is not the un- changed afferent impulse directed into a descending path, but the product of afferent impulses just received, combined with impulses liberated from their resting-places in the tissue of the brain." This much we are surely warranted in accepting ; and we have a right to infer that when the continuity of these con- nections is destroyed, interrupted, or structurally impaired, modification of function must ensue.

Although at present we can only surmise the physical basis of mental phenomena, it may be convenient to allude to the theories of such investigators as Meynert and Ramon y Cajal as to the afferent and efferent paths. Meynert held that the superficial layer receives the afferent or sensory fibres, that the efferent lies through the apical processes of the pyramids, and that the fusiform cells of the inner zone (his fifth layer) exercise, as association media, a trophic function. When Meynert wrote, Cajal had not demonstrated the large cells of the superficial layer, or his system of terminal fibres, and, apparently, the first recorded observer never expressed an opinion as to the origin or distribution of afferent nerves. Cajal's fourth system of fibres running directly to the superficial layer without passing through any cellular structure where they are in contact, but not anastomotic, with the "splendid tufts" of the pyramids, supplies this defect in anatomical connection. They complete the arc. Every physiologist must yearn for full confirmation of the existence of these organs. This author goes so far as to say that with some limitations it may be affirmed that the psychical functions are inseparably associated with the presence of the pyramidal, or, as he calls them, "psychical" cells. Apart from other considerations he holds this view on the ground that as we ascend in the animal series these bodies become larger and larger and more complicated. Further, he is of opinion that the psychical cells may exercise their functions more fully and more usefully the greater the number of the collateral expansions of their axis-cylinder, and the more copious, broad, and ramified their lateral and basal expansions, which, as shown by Golgi, have no anatomical connections, and must therefore transmit impressions by contact.[1] Golgi held that the protoplasmic processes exercised a nutritive function in connection with the cell, but he adduces no evidence sufficiently strong to support the theory.

Such is the line of thought and research amongst a large class of anatomico-physiological workers of the present day; and although the student should always maintain a neutral attitude of doubt, he cannot but perceive that certain of the mysteries of the brain's labyrinth are being gradually dissipated. But let it be remembered that the most earnest adherent of this school only speaks of his anatomical demonstrations as affording indications of data for certain physical conditions as necessary for psychical acts. He does not attempt to throw any light on the nature of these acts, which appear at present to be outside physiological theory. No hypothesis can at present explain how a peripheral excitement having reached the first cerebral layer can be converted into two such different things as a motor act and a state of consciousness. No one attempts to go further than to try to demonstrate the paths of conduction and dissemination.

[1] *Nuevo Concepto de la Histología de los Centros Nerviosos.*

CAUSES OF IMPLICATION OF THE APPARATUS.

It might be interesting to endeavour to range all the causes of Insanity—*i.e.*, causes of solutions of continuity—under the five comprehensive etiological categories set forth by Rindfleisch as applicable to disease at large,—over-exertion, injury from without, parasitism, deficient rudiments and defective growth, and premature involution or obolescence. The inquiry may, however, be conveniently deferred to a later part of the course. It may be sufficient to state now that these categories may be held to comprise a large proportion of the insanities, and that with certain additions, and an elaborate system of sub-classes, the whole could be comprehended. In the meantime we will confine ourselves to a consideration of the first of these etiological groups

The study of the pathological results of over-exertion of the brain is, to my mind, at once the most interesting and important subject that can occupy our attention. The resultant mental phenomena constitute the most common and most easily cured form of Insanity ; their study opens up channels of inquiry as to the action of all other morbid influences ; and on a fair estimate of the connection of this cause with its effects depends the position to be taken by the subject at large in the mind of the medical profession.

Am I doing the slightest injustice to any author of any work on so-called psychiatric medicine when I say that all attempts to relate what are generally spoken of as "moral" causes with insanity take the form of endeavours to construct a psychological nexus between cause and symptom, without demonstration of the intervention of any structural change in the cortical tissues. In a former course of Morison Lectures I endeavoured to show cause why we should regard these "moral" causes as inducers of physical changes in the cortex which may result in morbid phenomena of varying intensity. The arguments then adduced have (I hope) been amended since in various contributions ; and especially in the article "Pathology" in the *Dictionary of Psychological Medicine*, written in collaboration with Sims Woodhead. It is not my intention to reiterate all the statements therein made *in extenso ;* but in order to pre-

C

serve continuity it will be necessary to sketch out the general argument, supplementing it by the production of certain recent observations bearing on the general question.

There can be no doubt that hyperæmia of the superior surface of the brain occurs in direct relation to psychical activity. Numerous instances are cited by Mosso and others. I have observed the phenomenon in two subjects, and George Gibson has recorded tracings taken in a case in which the brain bulged through a hole in the skull during mental action, and receded as stimulus was withdrawn. It is still an open question how this functional hyperæmia is produced,—whether by reflex inhibition of the vaso-constrictor centre, by direct action of vaso-dilator fibres, or by a combination of the action of the two systems. It may be freely admitted that there is no anatomical evidence of the existence of such nerves, but physio-logical and anatomical considerations tend to show that one or both must exist. In the first place, Mosso has conclusively shown that during psychical activity considerable increase of the general blood-pressure occurs ; in the second, we have the fact, first stated by Moxon, that the greater veins of the pia enter the longitudinal sinus at such an angle that, as the blood enters the sinus, it is directed against the general backward-running current, so that unless the blood in the sinus is much diminished in quantity there is always retardation of the venous flow, and maintenance of a mild mechanical congestion. These two conditions must favour the rapid production of hyperæmia, and the existence of some controlling nervous influence may almost be inferred.

Bevan Lewis, starting from the fact that the blood conveyed by the two vertebral arteries is poured into the basilar—the sectional area of the single vessel being but slightly, if at all, greater than that of the two vessels going to form it—concluded that a "sustained pressure of no inconsiderable degree" is kept up in the minute nutrient arteries passing direct from it into the pons. Accepting this statement too readily, I suggested that this sustained pressure may not be entirely expended on the vessels of the pons, but may extend to the arteries given off by the circle of Willis. Lately, however, I began to doubt the accuracy of this opinion. On discussing the subject with

Dr Harry Rainy, he undertook to prepare a scheme of the encephalic circulation which would test the position. By the employment of india-rubber tubes and glass T-tubes the conditions were produced. Manometers were placed on the vertebrals, basilar, carotids, and circle of Willis, and tracings taken on a drum. The conclusion come to was that the original statement was inaccurate. The excursion of the pulse is certainly greater in the basilar than in the vertebrals, the excess of amplitude varying, *inter alia*, with the varying resistance to the egress of the blood from the larger vessels arising from the circle of Willis. But the mean sustained pressure is always lower in the basilar than in either vertebral, except in such cases as may be hypothetically supposed, when, both vertebrals being occluded, the blood current flows backwards from the carotids. We may therefore discard this hypothesis from the list of inducers of hyperæmia.

It is worthy of note that Rainy's experiments showed that variations in supply from the carotids alter not only the average pressure of blood in the whole system of supply to the base of the brain, but also directly influence the *locus* of maximum pulse wave.

Roy and Sherrington introduced acids and acid brain filtrates into the cerebral circulation, and found that hyperæmia was the direct consequence. Keeping before them the fact of the absence of anatomical proof of the existence of vaso-motor systems in the organ, they came to the conclusion that "the chemical products of cerebral metabolism contained in the lymph which bathes the walls of the arterioles of the brain can cause variations of the calibre of the cerebral vessels ; that in this reaction the brain possesses an intrinsic mechanism by which its vascular supply can be varied locally in correspondence with local variations of functional activity." [1]

The observations of Langendorf and Gescheidlen appear to be conclusive as to the alkaline reaction of normal brain tissue, and the rapid production of acidity under abnormal conditions, such as the interruption of blood current by pressure on the carotids. Keeping in view the physiological and anatomical observations stated above, it appears to me that the strong

[1] *Journal of Physiology*, vol. xi.

probability is that acid products may so increase the irritability of the muscular wall of the vessels as to render it all the more susceptible to vaso-dilator or vaso-constrictor influence.

Dr Milne Murray has devised apparatus by which the rise of temperature incident on chemical changes produced by psychical activity can be demonstrated. The principle involved in this instrument is the well-known one that the electrical resistance of a metallic wire varies with the temperature within wide limits. If, accordingly, the resistance of such a wire be measured at two different temperatures, the difference between the two temperatures can also be estimated. To apply this principle, a short length of pure platinum wire is arranged in a convenient way, and applied, for example, to the temporal region, the ends of the wire being connected to a specially arranged Wheatstone's bridge, and balanced against a constant resistance. The brain is kept as quiet as possible, and when a balance is obtained, the subject of experiment is asked to read or to try to solve a problem in mental arithmetic. In a second or two the index of the galvanometer begins to swing to the side which indicates that the platinum "thermometer" is being heated. Moreover, the arrangement can be made differential by placing a "thermometer" on each temple and connecting them to the opposite arm of the bridge. A balance is obtained with the brain quiescent; but at once, on cerebral activity being started, the index begins to move, showing that one side of the cranium heats quicker than the other. In right-handed subjects the rise of temperature is usually greater on the left side.[1] [Dr Milne Murray demonstrated these phenomena on two subjects.]

Mosso has clearly demonstrated that fatigue caused by psychical action produces not only a poisonous effect on the general system, especially the muscular, of the person experimented on, but he has also shown that the injection of blood taken from an exhausted animal into the circulation of animals at rest produces in them the extreme indications of fatigue. He is of opinion that the symptoms of poisoning are produced more from the breaking down of nerve cells than from any impairment of nutrition.

[1] A full description of the Bridge employed will be found in the *Transactions of the Royal Scottish Society of Arts* for 1893.

This observation brings into prominence the experiments of Hodge. Following up the observations of Sadovski and Madame Ternowski, he made a long series of experiments on the effects of electrical stimulation on the spinal ganglion cells of frogs and certain warm-blooded animals, and on the effect of normal fatigue on the cerebral cells of the sparrow, pigeon, swallow, and honey-bee. In the case of electrical stimulation, a nerve going to one or more of the spinal ganglia on one side of the animal was acted on for certain lengths of time, sometimes continuously, at others at intervals. The animal was then killed ; the stimulated ganglion and the corresponding ganglion of the opposite side were removed, and submitted to identical treatment for microscopic observation. The effects of fatigue on birds and bees were observed by securing specimens of each genus morning and evening, and preparing the organs in the same manner. The results of these observations, as recorded, are very fairly regular and uniform, and are to the effect that metabolic changes in nerve cells are as easy of demonstration microscopically as those of gland cells, and that they are of the same character ; that in the nucleus, which is the seat of most active change, after stimulation or normal fatigue, there is a marked decrease in size,—a change from a smooth and rounded to a jagged, irregular outline, and loss of the open reticulated appearance, with a tendency to take on darker stains than the nucleus of the resting cell ; that in cell protoplasm there is a slight shrinking in size, with vacuolation in the cells of the spinal ganglia, and in the cerebral cells a considerable shrinkage, with enlargement of the pericellular lymph space of the cells of the cerebrum and cerebellum, and decreased power of taking on stains. The amount of loss of bulk, as stated in the various cells, is considerable ; in the spinal ganglia the nuclear volume-shrinkage, after electrical stimulation, amounts to from 12·5 to 40·9 per cent., according to the length of time and degree of stimulation ; in the cerebra of fatigued birds, to from 36 to 69·7 per cent. ; and in the antennal lobes of bees, from 9 to 75 per cent.[1]

Dr Hodge has kindly sent me specimens taken from the

---

[1] *Journal of Morphology*, Boston, U.S.A., vol. vii. No. 2, "A Microscopical Study of Changes Due to Functional Activity in Nerve Cells," by C. F. Hodge, Ph.D., Clark University, Mass., U.S.A.

ganglion of the posterior root of the cat, and from the brains of honey-bees. Studying these sections along with Dr Mann of the Physiological Department of the University of Edinburgh, whose knowledge of the various conditions of the nerve cell exceeds that of any observer whom I have had the honour of meeting, the following conclusions have been arrived at as to the sequence of events. In the unstimulated cell the nucleus stains lighter than the protoplasm ; the first effect of stimulation is to reduce the staining of both to the same degree of intensity ; as it is continued the nucleus darkens, but remains lighter than the protoplasm ; then the nucleus becomes distinctly darker, and begins to get deformed and crenated ; and eventually a condition is produced which may be spoken of as collapse, nucleus and protoplasm losing all power of taking on stains. The first condition, that of equal staining, can be accounted for by the using up of nutrient material as the organ is brought into action ; the third stage is probably that observed by myself and others in the brains of the insane, and spoken of by Bevan Lewis as part of the evidence of over-stimulation, " dark staining of protoplasm, nucleus, and branches ; "[1] and the fourth stage the condition of actual or imminent vacuolation.

It appears to me that the third stage may be regarded as the limit of functional activity changes. It had been reached after eighteen hours' continuous excitation ; and it is interesting to observe how thoroughly the pabulum of a nerve cell may be used up and yet leave the organ in a condition admitting of reconstruction. The fourth stage appears to indicate a condition from which it is doubtful whether full recuperation can be obtained.

More recently Lambert has experimented on the cervical ganglia of non-anæsthetised animals. He confirms the observations of Vas, that on the stimulated side the protoplasm is divided into two layers, which is not the case with that of the organ at rest. In other respects his results are not identical with those of Hodge, but he shows marked differences on the two sides. The relative value of the two sets of observations has yet to be determined.

If we correlate the results of Mosso's and Hodge's experi-

---

[1] *loc. cit.*, p. 476.

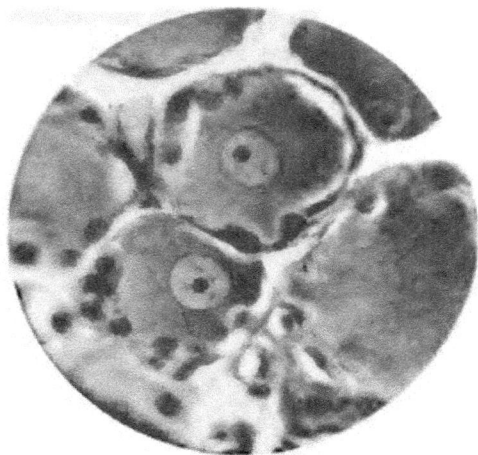

Unstimulated Cell from Posterior Spinal Ganglion of Cat.

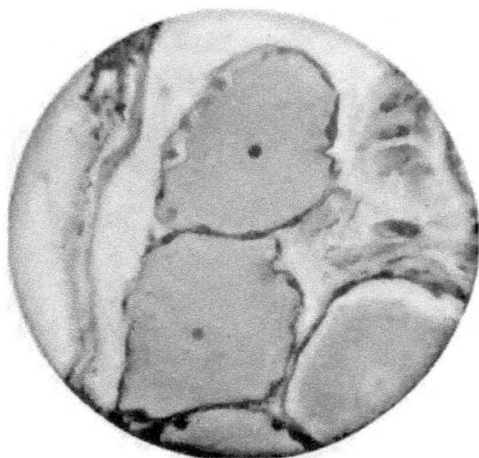

Cell from Posterior Spinal Ganglion of Cat. First effect of
Stimulation.

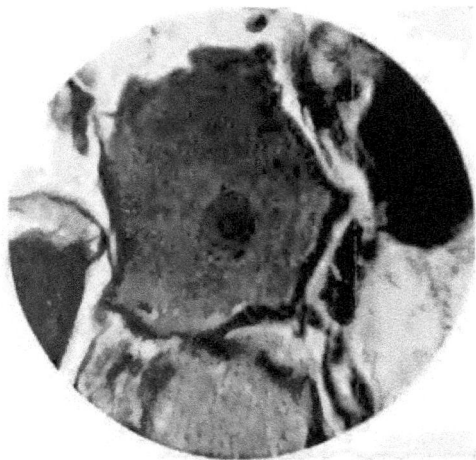

Cell from Posterior Spinal Ganglion of Cat. Second stage
of Stimulation.

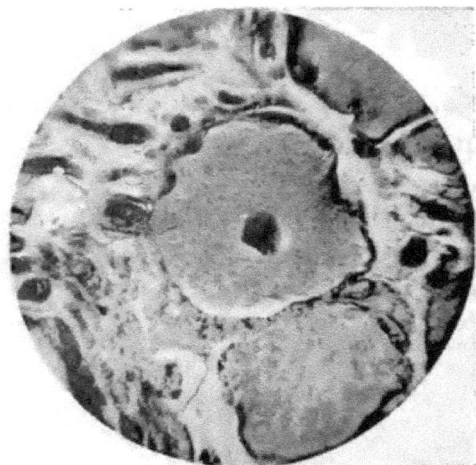

Cell from Posterior Spinal Ganglion of Cat. Later effect of
Stimulation.

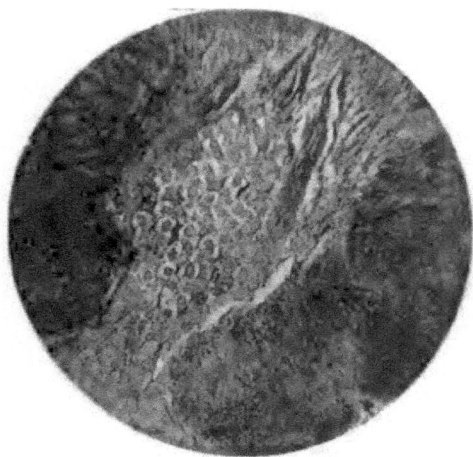

Sub-œsophageal Ganglion of Honey Bee.   Morning.

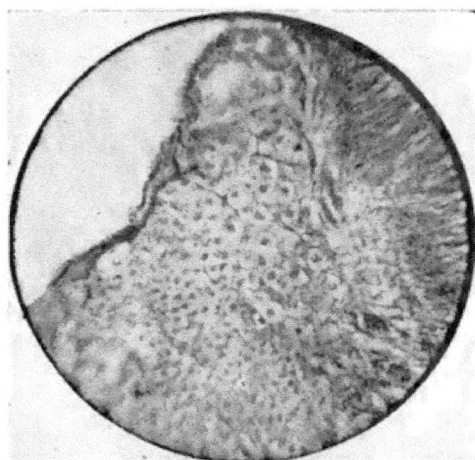

Sub-œsophageal Ganglion of Honey Bee.   Evening.

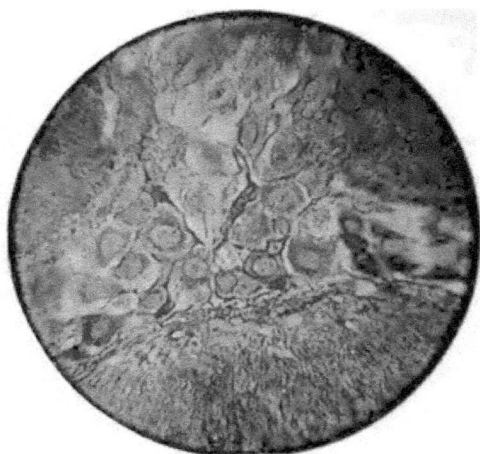

Antennal Lobe of Honey Bee. Morning.

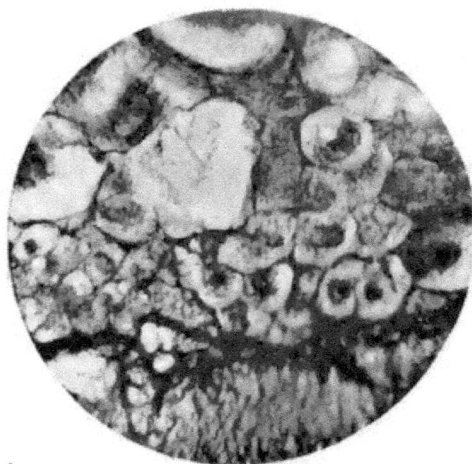

Antennal Lobe of Honey Bee. Evening.

ments we are led to the conclusion that the poison of fatigue is elaborated from the material lost during the period of work. Hodge's experiments extend to the process of recovery of individual nerve cells after electrical stimulation, which he finds to be slow, being scarcely complete after twenty-four hours' rest. It may be almost inferred from this that the degree and length of time of electrical stimulation applied exceeded the quantity which the cells are constructed to bear. The birds and bees he examined in the morning may be presumed to have done a previous day's work, and, so far as can be judged, their cells had reacquired functional integrity in twelve hours.

Halliburton and others have demonstrated the presence of albumoses in the cerebro-spinal fluid taken mostly from cases of spina-bifida. Such specimens cannot be regarded as healthy, and certainly cannot be used to illustrate the condition of the fluid after work ; nor do these observers give any opinion as to the nature of the albumoses. We have no evidence, therefore, as to what the poison actually is ; but I am informed by students of physiological chemistry that it is probably something midway between albumose and a body of simpler constitution, possibly a body closely allied to urea. All we can say at present is that it is of acid reaction.

With the possible exception noted, the experiments of Mosso and Hodge may be held to apply to the condition of organs under strictly physiological conditions. Expenditure of energy means expenditure of material, which is met by functional hyperæmia existing for such time as the stimulus is maintained. When it is withdrawn rest is obtained, and the vessels return to their ordinary calibre through constricting nervous action. But stimulation, abnormal in degree, kind, or duration, may make such demands on the local circulation as to produce in the brain, as in all other organs, pathological conditions not confined to the vessels themselves, but extending to the tissues they supply.

If we consider what occurs in other organs, such as the liver and kidneys, as a consequence of unphysiological hyperæmia, we find that a condition of unstable equilibrium between nutrition and function may be reached, productive of irregularity of action of their apparatus, and of implication of their vaso-control systems. In the brain we may have identical morbid processes going on. Continuous or frequently recurring excitation or irritation calls for an increase of blood ; the function of the vaso-motor nerves becomes overstrained ; the consequent dilation of the vessels is maintained by the lessened alkalinity of the cerebro-spinal fluid ; and the discharge of energy of the cells becomes irregular in consequence of the presence of more blood than is needed for repair, and the discharge takes place at a low level of cell nutrition and function. If this state of matters persists for any great length of time, and no relief is obtained, the sequence of pathological events common to all tissues under irritation ensues,—active hyperæmia, passive hyperæmia, and congestion. What is, to all intents and purposes, a sub-inflammatory condition is reached, evidenced by deposits of leucocytes, much greater than normal, between the hyaline membrane and the muscular coat, by red corpuscles in various stages of degradation, by large masses of pigment, by proliferation of the fixed connective tissue cells of the vessel, and by exudation.

Concurrently with, or at short periods after, the establishment of hyperæmia, we have implication of the important functions of the lymphatic system. On account of the basal position of the openings of the skull it is evident that even slight pathological alterations, either at the vertex or the base, but especially at the vertex, must implicate the maintenance of the vascular unity of the cerebrum, if they interfere with the removal of lymph fluid. Continued upward pressure must dam back the fluid by interfering with the patency of the pial conduits ; the only relief being found by the action of the Pacchionian villi. I need not tell you that in subjects of chronic insanity these organs are often found, after death, much hypertrophied ; this is doubtless due to the demands which had been made on them for the relief of the waste products of the organ. Doubtless this condition of the Pacchionian bodies is found in non-insane subjects ; but in those I have met with the brain history had not been ascertained. There is, however, no doubt but that the condition is much more common in the insane than the sane. But apart from this, obstruction to the outward flow may result from the deposit of *débris* produced by the breaking down of leucocytes, endothelium, and by deposits of blood-pigment, which are apt to collect in the perivascular lymph space. Collections of such material can be seen in large quantities within the canals, which can often be traced in squeezed-out specimens by its accumulation. The drainage being thus obstructed, the cells are subjected to the action of a fluid which is not only as to reaction in an unphysiological condition, but is also loaded with the products of waste. The pericellular sac is connected with the hyaline sheath by a fine tubule, the aperture of which is apt to be occluded by an abnormal flow of exudate carrying with it the *débris* above spoken of. And, again, such exudation fluid, when confined, probably acts detrimentally, by the exercise of pressure, on the conducting power of the vaso-motor systems, and thus helps to maintain the general condition of hyperæmia and congestion. In evidence of these theories, or, as I believe them to be, facts, I submit to you specimens taken from the convolutions of four patients who died within two months of the incidence of insanity ; in two of these the symptoms were those of mania, and in the others those of excited melancholia. The specimens illustrate the conditions I have been speaking of, and the differences they exhibit are more in degree than in kind. It is not often one has the opportunity of examining such subjects, for death does not usually ensue : it is only occasionally that such cases die from exhaustion, suicide, or some intercurrent disease, pneumonia being the most common cause of fatal results. Nor do I maintain that in all cases you have the same intense implications as are shown in these specimens ; still I think we have every right to assume that cases minor in degree as to symptoms are dependent on morbid action and pathological products similar in kind, although less in degree.

It is not my intention to trace the further solutions of continuity of the brain-elements produced by congestion of the convolutions, for those now submitted to your consideration are, I believe, the causes of the early symptoms of Idiopathic Insanity,—*i.e.*, the insanity produced by over-exertion of brain function.

The initial symptoms of this disease are generally insidious in character. Founding on the statements made by patients

D

suffering from premonitory symptoms, on those made by others who, after recovery from acute attacks, are able to carry back their recollections to the incidents of the prodromal stage, and on direct observation, I have no difficulty in asserting that physical phenomena are the first to present themselves. These consist of a feeling of fulness or uneasiness of the head, or of a dull, heavy pain extending from the forehead to the vertex, or even further back to the occiput. In some cases throbbing of the temples and eyeballs is complained of. Along with these sensations there is a general feeling of *malaise*. The pulse is generally irritable, not much increased in rapidity, but fuller than normal, and it may gain so much in volume as to assume a " cerebral " character. The first heart-sound is often accentuated. The temperature rises slightly at night. The general system becomes impaired ; disturbances of the digestive organs manifest themselves, and the nutrition of the body suffers. Oxalates or phosphates are usually present in the urine. In women the function of menstruation is affected. The mental symptoms follow closely. They consist in anxiety, restlessness, irritability, inability to apply the mind to the ordinary affairs of life, a strong tendency to introspection and concentration on self, and sleeplessness. Up to this point it is rarely possible to predict, with any certainty, whether the case will culminate in acute mania or in acute melancholia. If, however, no check is applied to the progress of morbid processes, loss of the function of control over ideas is manifested by excitement in one of two directions—(1) by the domination of one set of ideas, which are mostly of a depressed character ; or (2) by a tendency to follow lines of thought suggested by accidental circumstances,—that is to say, either a condition of acute melancholia or of acute mania is established. I have spoken of these two conditions as states of excitement, for we must bear in mind that although acute melancholia is manifested by depression of feeling, and acute mania by exaltation of feeling, they are both manifestations of excitement of feeling. Further, it must be remembered that although in many cases the character of depression or exaltation is maintained throughout, a large class exists in which it is impossible to say whether they are melancholic maniacs or maniacal melancholics.

This chain of symptoms is, I submit, distinctly referable to the brain conditions which have been sketched. They are the outward expression of gradually increasing fatigue affecting the whole system, and are manifested in the brain by what appear at first to be slight, but what are really important implications of psychical function. Along with abnormal hyperæmia we have headache, irritability, and restlessness. In connexion with the last symptom, restlessness, it may be well to remind you that the naked-eye evidence of morbid action in the insane is focussed around the so-called sensory-motor areas. In the very large majority of all cases dying insane we find the milky opacity of the arachno- and visceral pia, which is closely associated with underlying morbid processes, present in a space which can be covered by the two hands placed together, the lower ends of the hyperthenar eminences covering the spot where the fissures of Rolando meet. Is it pressing theory too far to suggest that, as the cells in this region of the cortex being the first to meet with morbid exudates, and being those of whose functions one appears to be the transformation of sensory impulses into motor acts, restlessness may be the indication of impairment of their activity? The giant pyramids are the first to show marked altered structure, and it is for this and other reasons difficult to refuse the inference that they are the structures implicated, and that, concurrently with the advance of patho-logical events in them, what was at first mere restlessness waxes into the excited action of acute mania or acute melancholia.

I think that by certain expressions I have indicated a belief that what are apparently widely divergent morbid mental symptoms—mania and melancholia—are dependent on common pathological conditions. The following clinical observations support the position as to the unity of pathological causation :—

1. During the prodromal period the symptoms of excitement and depression alternate.

2. In many acute cases mania and melancholia co-exist.

3. As many cases run their course towards recovery the symptoms are consecutively melancholia, mania, and dementia.

4. In *folie circulaire* the same sequence of symptoms occurs time after time.

5. In general paralysis of the insane a certain proportion of

cases is characterised by exaltation of feeling, another by depression, and a third by obfuscation of intellect, from beginning to end ; whilst in certain others we may have all varieties and degrees of symptoms.

6. The effect of the administration of certain poisons, especially alcohol, is a sequence of psychical phenomena of much the same character.

These observations point, not to a difference in kind of primary causation, but to variations of symptoms in accordance with the progress and nature of pathological processes, which vary principally in accordance with the constitution of the tissue of the individual. It must be borne in mind that the deposits of inflammatory products and of congestion are not identical or constant in the individuals of a series of subjects, because the individuals and their tissues are not constant quantities. We have, then, a constant condition of irritation acting on inconstant subjects. Given this constant psychological condition of excitement, whether it be manifested by exaltation or depression of feeling, we must seek for an explanation of the varieties of its phenomena in some quality or quantity of its exciting cause, in some peculiarity of its pathological products, or in some idiosyncrasy of the affected individual. We derive no material assistance from psychological considerations, for there is no necessary connection between depressing emotions and melancholia, on the one hand, or between stimulating emotions and mania on the other. Intense grief produces mania as often as melancholia, and the insanity of the man of saturnine disposition is as often as not characterised by mania. The peculiarity of the exciting cause appears to be, not its psychological characteristics, but its intensity and rapidity of incidence, the latter depending not only on the former, but also on the stability or instability of tissue. According as excitement of feeling is rapidly produced, so the more likely is mania to be the symptom, in consequence of the action of abnormal arterial hyperæmia, and its immediate consequences. I have before my mind's eye the case of a man who, on hearing by telegraph of the death of his wife, rapidly developed symptoms of an exalted maniacal character. It is not only to the construction of the cortical cells and their network to which we may look for evidence of instability and irritability, but also to the ganglia which govern the vaso-motor systems of nerves. Inherent weakness of these centres may play an important primary *rôle* in the production of insanity, more especially in relation to the rapidity of its production. That melancholia often supervenes on depressing emotions, gradual in their incidence and action, does not imply a psychological nexus, but that as their irritating influence is generally slowly applied to cells of diminished vitality and nutritive power, so the results of irritation are slowly produced ; and, as in the case of every organ of the body, we have variety of degree of symptoms in conformity with the rapidity of the progress of pathological events. In extreme cases of recent excitement, maniacal and melancholic, I have found stasis and the results of inflammation of a subacute character ; in chronic cases the same appearances are presented, although in a less degree, whatever the symptoms may have been ; and if we have any warrant to connect post-mortem demonstration with the indications of disease during life, the inference is unavoidable that considerable variety of clinical phenomena may be dependent on a common cause acting on differently constituted tissues.

LECTURE III.

In the last lecture I directed your attention to the fact that the naked-eye indications of diseased action in the brains of those dying insane are, as a rule, confined to an area corresponding in extent to that of the two hands placed side by side on the vertex. At the base the evidence of morbid change is, for the most part, absent: the focus of diseased action is the line of the Pacchionian villi and the Rolandic area; and its indications fade away anteriorly, posteriorly, and laterally, till they merge into the unaffected pia of the more dependent and inferior regions. It may be accepted as a fact that the condition of the pia, as viewed by the naked eye, affords a fair index of the degree of *active* morbid action going on in the subjacent convolutions; microscopic examination confirms the statement. How is it, then, that the areas which we are taught to believe form, in the words of Ferrier, " the substratum of those psychical processes which lie at the foundation of the higher intellectual operations,"[1] and the aberration of whose functions is the prominent symptom of the disease we are considering, present in most cases slight visible departures from normality of the membrane; in a certain proportion, none. This question is difficult of solution, but I think the attempt to work it out may be productive of points of interest bearing on the general position.

The focus of diseased action—*i.e.*, the ascending frontal and parietal convolutions, and the neighbouring parts of the superior and middle frontal gyri—sends downwards from certain of its zones of cells (according to Ramon y Cajal, the pyramids) axis-cylinders which are in connection with fibres forming the pyramidal tract, which can be traced through the inner capsule to the crusta. It is unnecessary to remind you of the connections of

[1] *The Functions of the Brain,* p. 147.

this tract with the skeletal muscles, and with the cranial nerves for
the movements of the eyes, mouth, face, tongue, pharynx, and
larynx, which have been proved by direct stimulation, and by the
observation of downward degenerations.   I do not ask you to
accept without reserve the view that the centripetal or terminal
fibres of Ramon are the afferent paths.   Ramon does not affirm
the proposition ; he merely asks, " Is it possible that such
fibres represent the cerebral termination of sensory nerves, or at
least that of axis-cylinders proceeding from cells dovetailed with
the terminal branches of those nerves?   This seems probable,
but cannot be certainly ascertained."   However this may be,
no one can doubt that some such executive afferent apparatus
exists.

In this connection I wish to draw your attention to an
interesting and suggestive observation made by Groves.[1]   His
experiments were directed to the effect of chemical stimulation
on motor nerves.   In summarising his results he states (*inter
alia*), that "the present state of our knowledge in regard to
chemical stimuli leads to the conclusion that the nerve impulse
which connects the causal stimulus with the resulting contraction
is of quite a different nature in the cases of chemical and electrical
stimuli ;" and that "afferent nerves are almost unaffected by
chemical stimuli."   This author dealt only with peripheral nerves
of the extremities and the vagus, and we can only use his conclusions by analogy : but if it is admitted that brain exudates
exercise an influence on cell or fibre, it is evident much interest
attaches to the observations.   If the afferent apparatus is, and
the efferent is not, affected by its immediate surroundings, serious
modifications of function would appear to be inevitable.

Every discovery appears to bring to light important facts
bearing on the anatomical association of adjacent and remote
parts of the brain.   The *rete mirabile* of cell and process fibrils is
almost inconceivable.   We have the collateral fibres from axis-
cylinders and terminals, the arborisations of the protoplasmic
processes of the pyramidal layers, the ascending processes of the
cells of the fourth layer, the collaterals of callosal, projection, and
association fibres, and the ramifications of the large white fibres.
Intricate as this network is in all regions of the cortex, it reaches

---

[1] " Chemical Stimulation of Nerves," *Journal of Physiology*, vol. xiv. p. 231.

its maximum in the superficial layer. Let it be remembered, also, that the association system of fibres of the white matter forms the great mass of the *centrum ovale*,—a suggestive fact not sufficiently impressed on the student in anatomical and physiological works ; that the axis-cylinders with which they are connected bifurcate, one of them passing back into the cortex, and the other going to the corpus callosum, while the true association fibres send off at right angles numerous ascending collaterals. Thus by every nervous structure, and possibly by the special action of the fourth layer, association is maintained between cell and cell, between cell and fibre, between convolution and convolution, and between hemisphere and hemisphere.

There are the very strongest reasons for believing that kinæsthesis (transmutation of sensory stimulus into motion) occurs in the pyramidal cells of the Rolandic area. This cannot, however, be the only transmutation of energy performed in this area ; important and essential impulses must be transmitted from the pyramids to cells of other regions, where they are conserved or released according to circumstances. Let me add the concluding words of the paragraph just quoted from Ferrier's work :—" There are centres for special forms of sensation and ideation, and centres for special motor activities and acquisitions, in response to, and in association with, the activity of sensory centres ; and these in their respective cohesions, actions, and interactions, form the substrata of mental operations in all their aspects and in all their range." [1]

The processes by which a sensation (using the term in its physiological sense without any psychological connotation) becomes transformed into (let us say) a sentiment, are outside physiological theory. Without prejudice to the opinion that nerve force is a physical force, it may be pointed out that its transformation stands on the same platform as the doctrine of the transformation of physical forces : both can be demonstrated experimentally, but the condition of matter incident on the processes is in either case subject of speculation.

The organs of the Rolandic area are probably not the only transmutators of energy, but it is evident that their action must

---

[1] *The Functions of the Brain*, p. 147.

have a wide-spreading influence on that of other regions. This area is the most active region of the brain, and, therefore, the most liable to accident. Given that its cells are affected by any evil influence (in the case we are considering, by the influence of over-exertion), we have a right to look to it for the effects of impaired transmutation. So far as we can at present judge, the pyramids are the first cells to show morbid changes. In almost all subjects we find evidence of work in these organs ; but the evidence of work due to functional activity increases proportionately with age and cause of death. I have examined the pyramidal cells of many animals and many human beings. In the former the indications of waste due to exercise of function, as manifested by pigmentation, are slight,—only such as we may assume are physiological in amount. In the young man or woman dying of acute disease, the same holds true ; as we take cases more advanced in life they increase, but not beyond a certain limit ; in those taken from subjects who have died after long, painful, or wearing disease, who have laboured under delirium or other head affections, or who have died insane, the indications of waste and of structural change become more and more pronounced, and the increase occurs in the order in which the causes of death have just been stated. In the large majority of cases in which morbid mental symptoms have presented themselves, and in which excessive pigmentation has been observed, the body of the cell undergoes changes in form, and its protoplasm, judging from the manner in which it takes on stains, also becomes changed in quality. The observations of Miles on the effects of traumatism on the convolutions[1] support my view ; and it is of great interest to note that in two human cases related by him this appearance was found to have occurred within fifty and twenty-five hours respectively of the infliction of the injury. The mere prominence of these cells by reason of their size does not in any way account for the observation ; for those of other layers have been carefully examined, and, as regards period and degree of implication, have not been noticed undergoing degeneration in the same rapid and intense manner as the giant cells. Certainly, in the cases which I have examined which had succumbed earliest after the incidence of insanity,

[1] *Royal College of Physicians of Edinburgh Laboratory Reports*, vol. iv. pp. 181-3.

none of the cells of the anterior two-thirds of the frontal convolutions and of the posterior occipital gyri were so deeply
implicated as those of the Rolandic area ; they were not nearly
so deeply pigmented, and their degeneration was of the granular
character.

Given a zone of important cells undergoing stimulation beyond
the physiological limit, we may expect effects to be transmitted
to remote organs exercising different functions, by means of the
intimate anatomical connections which have been demonstrated :
and it does not follow that these effects must be of the same
nature as those acting on the zone of cells primarily affected.
In the convolutions anterior to the Rolandic area the effects of
sub-inflammatory action are much less well marked in insane
subjects than at the vertex ; they are often absent, and the
pigmentation of cells is generally slight.   In this region the cell
degeneration is of a granular character, which may more or less
rapidly go on to destruction and absorption.   This, to my mind,
is indicative of impaired nourishment rather than of active
morbid action,—of interruption of trophesial function, which
depends on the maintenance of the integrity of cell connection.
To this destruction of cell material and of the implication of
nerve fibre, presently to be spoken of, must be ascribed the
general shrinkage of the anterior lobes so often noted in the
insane.   To illustrate its frequency I quote from Bevan Lewis :
—" The fronto-parietal segment of the hemispheres enjoys least
immunity from the atrophy incident on insanity, and we find
atrophy confined to the frontal lobes in a large number of cases—
in fact, in the proportion of three-fourths to one-fourth respectively." [1]   In the course of almost every case of idiopathic
insanity—*i.e.*, insanity due to over-exertion of the brain—we have
a fairly well-marked prodromal period indicating the diseased
balance between nutrition and function in the kinæsthetic area.
As the impaired function of the cells of this area makes itself
felt on the cells of ideational centres, whether by lessened or
increased conduction of energy or by modification of their
nutritive influences, we have reduction of their structural integrity, and perversion of psychical acts.   As every case runs its
course to recovery or to permanent insanity, we have a period of

[1] *Loc. cit.*, p. 457.

mental and physical torpor, which we speak of as dementia. On the rapid removal or non-removal of the causes of implication of the cells of the Rolandic area depends the issue of the case— recovery or prefrontal atrophy.

The comparatively slight milkiness of the frontal pia as contrasted with that of the Rolandic area is dependent, not on underlying morbid action, but on the damming back of the lymphatic stream, which, running upwards, is arrested at the vertex in the line of the Pacchionian villi, and the *débouchement* of the great veins. We all know that this appearance presents itself in most subjects, sane or insane, and that its intensity increases with age. But, except in the case of inebriates, it is never found so well marked as in the insane, and in their case it is prominently indicated at a much earlier period of life than in the sane. Its first appearance is as a fine milky line along the course of the pial veins, and is due to the deposit of waste and plastic exudates. As these accumulate and diffuse, the membrane becomes thick, tough, and on section is found to consist of a mass of material which looks like an immense increase of the normal trabeculæ. Through this thickened structure the flow is slow, and the waste material which would have been disposed of were the channels patent is deposited in a certain circumscribed area. It is probable that the fluid is filtered in the thickened pia, and that as it reaches the larger cisterns it is so slightly loaded as to pass freely into the wide space in the spinal canal. What anatomical limitation governs this diffusion? It is an interesting fact that, except at the vertex, the widest stretch of the milky area corresponds with the distribution of the branches of the middle cerebral artery ; the rest of the arachno-pia (except in such acute conditions as general paralysis) being for the most part clean. Is it that the demand for blood by the active part of the organ leaves after it the signs of work and the evidence of waste of tissue?

But this suggestion does not meet the fact that at the vertex on either side of the sinus, and especially in the neighbourhood of the Pacchionian villi, the grossest thickening and opacity of the membrane is most frequently and markedly present. This area is generally supplied by the anterior cerebral artery ;[1] on

---

[1] *Vide Quain's Anatomy*, tenth edition, 1893, vol. iii. part 1, p. 196.

the mesial aspect we have no sign of disease ; why, then, in this region ?   In the first place, it is the area of *débouchement* of the great veins ; and, secondly, it seems highly probable that the Pacchionian villi perform a much more active lymphatic function than they generally get credit for.   Should extra work be laid on them by failure of the other lymphatic system they develop in size ; notwithstanding which, not being able to meet the strain, a proportion of their plastic contents escapes, and affects contiguous parts.

As yet I have only alluded to the effects of unphysiological exudates on vessels and cells ; their effects on nerve fibres and cell processes, although they may be of secondary origin, are still of very considerable importance.   It is well known that, so far as experiment can show, it is impossible to tire a nerve fibre, and that it will continue to transmit impulses under very unfavourable conditions.   Also we know that in the brains of old-standing cases of insanity marked departures from normality are found in this tissue, and that they are especially distinct in the frontal lobes. From my point of view one of the most important of Miles's observations is that we have evidence that changes in nerve fibre, and probably in the collaterals, and in other processes of cells, can be very rapidly produced, and are indicated by the presence of colloid bodies and so-called " miliary sclerosis."   The latter lesion was described by me many years ago, along with Prof. Ruther ford, and I am mainly to blame for the error enunciated as to its genesis and name.   Bevan Lewis convinced me that colloid bodies and so-called " miliary sclerosis " were aggregations of myelin derived from the investment of nerve fibre.   I found the former in the brains of birds within ten days after the receipt of injury ; but Miles discovered it in the case of a boy who died fourteen hours after an accident, and in a man who survived his injuries only eighteen hours.[1]   In animals he found it very rapidly produced after injury to the head.   Miliary sclerosis is merely the result of the confluence of larger droplets of myelin. I am, of course, very far from saying that these evidences of nerve lesion are produced in anything like the same short space of time in the brains of insane persons ; these observations are only adduced to show that they may be present in cases much

[1] Miles, *loc. cit.*

earlier than could be inferred from any extant statement. In Miles's cases these bodies were doubtless produced by bruise and rupture of the fibres. Bevan Lewis holds that they are "accidents occurring in the course of a sub-inflammatory or degenerative change in the medullated nerve tracts."[1] It is quite possible to conceive that in the earlier stages of the condition the droplets of myelin are very small, and that they do not coalesce till lacunæ are formed by the further action of disease. We are all aware of the effects of certain poisons, such as the primary infective agent of typhoid fever, tubercle, beri-beri, and diphtheria, and of alcohol, on peripheral nerves. The medullary substance is first affected, and becomes attenuated; this is followed by interruption of the axis-cylinder, and, as a consequence, degeneration of the nerve throughout its course ensues.[2] It is a moot point whether these changes are of central or peripheral origin; for my present purpose it is un-necessary to debate the matter. My conviction is that, in the brain, interruption of connection is brought about by Wallerian degeneration of fibres, processes, and collaterals, the result of structural implication of cells produced by over-stimulation, and maintained and aggravated by the effects of poisonous exudates.

I have sketched—roughly enough it must be admitted—the general structure of a convolution, certain physiological con-siderations bearing on its method of conduction of impulses, and certain pathological conditions which may be productive of temporary or permanent perversion of one or more of its functions. Many interesting and important points have been omitted, as time would not permit of their being taken up in detail; but I submit enough evidence has been adduced to illustrate the existence of an apparatus of the most complex kind. What strikes one forcibly as one works at the subject is, if on the maintenance of this most delicate and intricate machinery in its entirety depends the healthy working of the body and the normal exercise of the mental functions, how is it that pathological and other accidents do not occur more frequently than they do? To compare the terminal fibrillæ of an axis-cylinder with a filament of a spider's web would be

---

[1] *Loc. cit.*, p. 264.
[2] *Vide* "Multiple Sclerosis," Sidney Martin, *Journal of Physiology*, vol. i., No. 3.

to compare a silken thread to a hawser. There is no organ of the body possessing such immense intricacy of arrangement as a convolution of the brain ; the structure of the spinal cord is, in comparison, simplicity itself,—that of the cerebellum or of the retina commonplace. It is impossible to find terms or to cite comparisons to express the feeling produced by study of its delicacy and complexity. It is preserved from rough accident by its rigid bony envelope, which has till of late years been regarded as " a practically closed sac." Were it so, it would prove a greater source of danger than of protection : a fit of coughing or of sneezing might under such circumstances prove fatal. If we assume that the average compressibility of brain matter is at all comparable with that of water, were it confined in a rigid chamber absolutely filled, the addition of a $\frac{1}{1000}$ part to its volume would produce a pressure of one atmosphere, and an additional atmosphere would crush the organ into a shapeless mass. The statement of the physical fact enables us all the better to recognise that the coherence and continuity of this large and delicate organ depends on the maintenance of the normal ebb and flow of the cerebro-spinal fluid in which it is suspended. The constant variation of the quantity of this fluid is admirably illustrated by the experiments of Dean,[1] which also enable us to get a clearer conception of its action in preserving the unity of the whole encephale. The cerebro-spinal fluid also receives from the lymphatic system the products of waste, and under abnormal conditions the products of diseased action. Any interference, therefore, with its flow must act detrimentally ; for, be it remembered, the brain obtains no vicarious assistance from other systems,—it must do its own work, and must rid itself of all effete matter.

Although Falkenheim and Naunyn[2] have shown that increase of blood pressure does not increase the amount of secretion of cerebro-spinal fluid, Dean observes that the cerebro-spinal fluid pressure increases considerably with increase of blood pressure ; and he naturally infers that the same occurs under such circumstances as excessive secretion or diminished absorption of the fluid, or the pouring out of inflammatory exudation.[3]

[1] "Cerebro-spinal Pressure," W. C. Dean, *Journal of Pathology*, vol. i., No. 1.
[2] *Archiv für Exper. Pathol.*, 1887.   [3] *Loc. cit.*

Further, he shows that local pressure produces œdema of contiguous parts, and that in the area submitted to pressure degeneration of cells takes place. I think that, given functional hyperæmia passing into congestion and subacute inflammation, we have every reason to believe that the lymphatic flow is interfered with, and that the brain must suffer in respect of functional activity and solutions of continuity of its structure. No one reading Bevan Lewis's chapters on Pathology can fail to be struck by the frequent reference to inflammatory changes affecting bone, membranes, and nerve fibres. Let us take his remarks on the pia :—" The pia mater is abnormally thickened in fully 48 per cent of those dying insane, partly from fibrinous exudates which have organised, partly from plastic lymph, and often from an œdematous swollen condition of the conjoined soft membranes."[1] In like manner he refers changes in the skull and dura to inflammation, and as we have seen, he speaks of colloid bodies and "miliary sclerosis" as accidents occurring in the course of sub-inflammatory action. I prefer to take this book because it is the most authoritative work on the subject in our language, notwithstanding that its author expresses no opinion as to the existence of sub-inflammatory changes in the early stages of "acute mania" and "acute melancholia." He generally speaks of the changes spoken of as being the result of old-standing inflammation : the meaning of which I infer to be that it occurred at an early stage. Certainly the 48 per cent. included general paralytics ; but that leaves a fair margin for cases of terminative dementia due to other causes of insanity. But in the presence of distended and congested vessels, leucocytes in the perivascular canals, and exudation, we have direct evidence of inflammatory action in all cases which had been the subjects of acute mania or acute melancholia.

[1] *Loc. cit.*

LECTURE IV.

WE must now turn our attention to a very important series of symptoms produced by morbid hyperæmia and congestion of the Rolandic area,—the impairment of the health of the general system. These somatic symptoms are of great interest, and their consideration has a very considerable bearing on treatment.

It is not my intention to discuss in any detail the vexed question as to the manner in which the central nervous system affects metabolism,—whether by the action of special "trophic" fibres, by the action of the vaso-motor systems, or by any undiscovered special quality of nerve regulating its government of muscle, gland, or connective tissue. The whole subject is far too complicated to admit of profitable discussion on the present occasion. We must content ourselves with the clinical fact that, under certain diseased conditions of the nervous centres and of their extensions, profound changes take place in the nutrition of the general system; that no tissue is immune from their influence; and that the functions of each and all suffer manifestly. We must work on the general principle "that the nutrition of each tissue is, in the complex animal body, so arranged to meet the constantly recurring influences brought to bear on it by the nervous system, that, when those influences are permanently withdrawn [or morbidly implicated], it is thrown out of equilibrium : its molecular processes, so to speak, run loose, since the bit has been removed from their mouths." [1]

The earliest symptoms of over-exertion of the brain have been already stated :—A general falling-off from the normal condition of the body ; the skin gets harsh or clammy, the complexion pale or dusky ; dry hair; uneasy, or actually painful, sensations are experienced at the vertex, less frequently over the

---

[1] *Text-Book of Physiology*, Foster, part ii.

temples and occipital region ; the temperature may be slightly raised at night, sometimes unilaterally ; the pulse is irritable, small, and compressible, increasing in volume as the case progresses, until it may become "cerebral" in character ; the first heart-sound is generally accentuated ; oxalates or phosphates are frequently present in the urine. Without any manifest disease of any particular organ, the whole system is out of tone. Apart altogether from the symptoms of restlessness, irritability, and anxiety, the patient is ill ; to use a common colloquialism, he is below par. His nourishing power is diminished. As a rule the digestive system is the first to suffer prominently, dyspepsia in various forms presenting itself. The muscular system next shows indications of loss of tone. This prodromal period may last for long unnoticed by relatives or friends, except so far as they may remark that the patient is not looking well, and is not quite himself. The patient himself is not unfrequently painfully conscious of the change, and consults his medical adviser as to his condition ; the symptoms most strongly weighing on him being the general sense of weakness, and the inability to apply himself to work. It is during this stage that we most frequently meet with abnormal salts in the urine. Are we to infer from this symptom that systemic degradation results from nerve waste, which in its turn may be ascribed to perverted metabolism of the superior cortical areas, and that reduction of nerve conductivity shows its effects on the glandular and muscular systems? To my mind it is impossible to refuse to draw the inference. But then the question arises—Is this a direct action, or is it through modifications of the vaso-control apparatus? I fear that in the present state of our knowledge the solution of this problem cannot be definitely arrived at ; but I venture to say that, keeping before us the general results of physiological and pathological inquiry at our command, the strong probability is that impairment of the vaso-control system consequent on nerve implication is the efficient factor in producing reduction of general functional activities. It appears to me that a rough analogy is afforded by the study of herpes and certain forms of scleroderma. In that condition we have an illustration of the coarse effects of impaired nervous control over limited areas, due to implication of ganglia ; and to more extended areas, due to

F

destructive disease of the cord and brain. According to Kaposi[1] the skin affection in herpes zoster appears to depend in many cases on a congested condition of the vessels supplying the ganglion or ganglia of the nerve or nerves supplying the parts affected. I only instance this disease as a possible illustration of the widespread effects on nutrition exercised by a localized curable nerve condition.

The blood suffers sooner or later in respect of loss of red corpuscles and a diminished amount of hæmoglobin. My experience is that this deterioration does not occur till towards the end of the prodromal period, or even later. It is not easy to get definite information on this point; patients are nervous, restless, anxious, and suspicious, and are apt to resent the necessary processes of examination. Moreover, the number of observations necessary to arrive at any thoroughly satisfactory conclusion is a further obstacle. In the few cases I have examined with any degree of satisfaction, I have failed to find definite indications of the quality of the blood becoming markedly affected at an early stage ; but as the case advances the blood certainly falls off in quality in the directions already indicated, the loss of hæmoglobin not being proportionate to the diminution of the number of corpuscles. There does not appear to be any loss of leucocytes. As to the relative amounts of the other blood constituents I have no experience ; and, so far as I can discover, no very authoritative observations on the subject have as yet been published. I cannot venture to generalize on the few cases observed by myself; they only support the general result of the careful preliminary work of Macphail[2] and Smyth,[3] that the blood in the insane is in a morbid condition ; but we have no scientific data to help us to arrive at conclusions as to this morbidity being a primary or secondary condition. Proceeding on clinical observation, we have strong reasons for believing that, in certain classes of insanity, the pathological condition of the blood is a direct consequence of impaired nervous influence acting on the blood-glands, producing results on the nourishment of the general system ; that, in others, mere deficiency in quality and quantity is the efficient factor in the

[1] *Pathologie und Therapie der Hautkrankheiten*, 3rd edition.
[2] *Journal of Mental Science*, vol. xxxii.   [3] *Ibid.*, vol. xxxvi.

production of morbid mental symptoms ; and, in a third section, toxæmia, using the word in a very broad sense, is the active cause. But the actions and reactions of the blood conditions produced demand careful inquiry. This is now one of the great fields open to the student of the pathology of insanity, who, in prosecuting his research, must not only address himself to analytical processes, but must employ experimental methods. Mosso has set the example by transfusing the blood of fatigued animals into the circulation of the healthy, and has obtained interesting results. No one will accuse me of the weak jocularity of suggesting that the blood of insane persons transferred into a healthy animal would produce similar mental symptoms ; this line of inquiry would probably throw light on the nature of blood degradation, and of that of any poisonous material it might contain, present in consequence of elaboration or of non-elimination.

As to my own rough observations made on the special class of cases we are considering, I can only say that they were taken for the most part during the prodromal stage and the early period of excitement. The general result was that blood changes were not manifested until actual insane excitement had appeared, and that in cases of active mania or melancholia the changes were not so marked as in those in which the mental symptoms were less acute. I need hardly remind you that the incidence of acute mania or acute melancholia (the term acute being employed as indicating intensity of excitement), is, as a rule, much more rapid than in the subacute conditions. Considering these statements alongside of the opinion expressed that oxalates and phosphates are found in the urine more abundantly at an early stage of the disease, it appears to me that the sequence of pathological events is as follows :—Acute congestion and nerve waste, productive of active symptoms, which, if not relieved, is followed more or less rapidly by weakened action of the glandular systems, and consequent deterioration of function of the alimentary canal and implication of the blood ; and as a condition of nervous congestion and stasis is reached we have to deal with a further reduced functional activity, not only of the organ primarily affected, but also of the various systems which it governs. Nerve waste has reached its minimum capacity for loss, as indicated by oxaluria and phosphaturia. Hodge's experiments have shown us how rapidly what I have termed cell-collapse can be reached : it is not difficult to realize what the consequences of the maintenance of that condition would be, or what disaster must follow if repair is not obtained.

Loss of tone of the digestive apparatus is an almost invariable prodromal symptom. It is manifested by obstinate constipation, large offensive stools, fœtid breath, eructations, furred tongue, and uneasy sensations. This condition is the objective starting-point for many delusions. It constitutes an important train of symptoms, whether we regard it from a physiologico-pathological or from a therapeutic point of view : for, as regards the former, it is another indication of the action of the diseased brain.

Accentuation of the first heart-sound, which is a common symptom, is of no importance, except so far as it is a further indication of the general implication of the nervous system. The same remark applies to the slight rise of temperature, which is seldom maintained for any great length of time.

Much greater importance is often attached to menstrual disturbances than is actually warrantable. Insanity is never directly due to arrest of menstruation. Its cessation is one of the symptoms of general degradation of the system, however caused, or of involution. Its return during convalescence does not mark a crisis in the case : it simply means that the reproductive organs are resuming function along with other systems. In that respect it is of importance, but in nothing more.

It may be well here, in connection with the bodily symptoms, to point out the dangers to accurate diagnosis of the too ready employment of an etiological classification of the insanities. In looking over a list of varieties of insanity, we find Amenorrhœal, Menstrual, and Post-menstrual Insanity enumerated, and we have the broad statement of such eminent authorities as Esquirol and Morel before us, that mental aberration is caused by menstrual disturbances in one-sixth of all cases.[1] I cannot help thinking that on its very face this statement carries the impress of too broad and precipitate generalization. No one will now accept it ; but it illustrates how prone the mind is to jump at conclusions as to the relation of cause to effect. We will all

[1] *Dictionary of Psychological Medicine*, vol. ii. p. 801.

acknowledge that in the insane menstrual disturbances occur very frequently, and that at the period the mental symptoms are liable to undergo various modifications, exacerbations, or diminutions. Further, it may be allowed that during the pro-dromal period of many insanities the monthly flow is arrested. But does it follow on these admissions that amenorrhœa is to be held to be the proximate cause of the mental symptoms, or that we have in a given case sufficient to warrant us in calling it one of " amenorrhœal " insanity. First let me appeal to your experience amongst women who from puberty have suffered from scanty or profuse menstruation, or dysmenorrhœa unassociated with disease of any sort. Have you, amongst such, seen insanity occurring as a direct consequence? There are many present whose experience in such matters must be much greater than mine; still, I can state that the large proportion of such women are, so far as I know, much the same mentally as the mass, and are not more liable to break down, unless the conditions of menstruation are associated with neuroses not necessarily accom-panied by insanity. Putting it conversely, have we not met with cases in which arrest was caused by severe mental shock, or has occurred during the progress of some acute bodily disease? When women reach the climacteric period, nervous symptoms, whether amounting to insanity or otherwise, do not present themselves at the exact period of menopause, but a year or so before the actual arrest, more frequently from one to two years after. Let us take a very common case—that of servant girls who, coming from country districts and farm-work to domestic service in a city, after a shorter or longer period fall off in health, cease to menstruate, and become depressed and melancholic. Are we warranted in assuming that such women are suffering from amenorrhœal insanity? I think not. The case, as roughly stated, gives us data for believing that both sets of symptoms, bodily and mental, are due to anæmia.

The great objection to the cataloguing of such cases under such headings,—the employment of incorrect diagnostic terms,—is that it may lead to errors in treatment. My conviction is that the weight of evidence favours the conclusion that in idiopathic insanity menstrual troubles are a consequence of the general impairment of the system, and that they are not important factors in the progress of a given case.   Pray do not mistake me : I do not deny that interesting phenomena occasionally present themselves in the chronic insane synchronously with menstruation, and that recrudescence and remission of symptoms may be noticed at the period.   All I desire to indicate is, that such phenomena are of no great import as regards the issue of the case, and only indicate that the nervous system is liable to stimulation from slight excitants.   Am I wrong when I say that the importance attached to the influence of menstruation on the course of a case of non-nervous disease is gradually waning, and that the physician does not watch its appearance with such a jealous eye as of old ; crises are not anticipated as concurrent effects ; and that the regularity or irregularity of the functional discharge is regarded more as an index of the general condition than as mysterious agents acting directly on the course of the case.

Another instance of mistaken relation between bodily conditions and mental symptoms is afforded in the case of diseases of the female organs of generation.   So far as the graver forms of disease, such as tumours, are concerned, I may say I have never met with a case in which insanity was thereby produced.   I have applied to many leading gynæcologists for their experience ; amongst others, Matthews Duncan and the Keiths, and their evidence is to the same effect.   Text-books on diseases of women are silent on the point.   That insanity may occur alongside of uterine or ovarian tumours is most true, as a result of anxiety, uneasiness, misery, and other inevitable accompaniments ; but that does not imply a peripheral irritative insanity, nor does the fact that in certain reputed cases of extirpation recovery of sanity has followed, prove any direct causative action of the tumour. For all practical purposes peripheral irritation may be dismissed from the list of producers of insanity.   Did it so act, the records of surgical hospitals would afford endless examples of its morbid action on brain health.

It is of importance to note the absence of insanity as an accompaniment or sequela of certain complaints which, *à priori*, might be supposed to be prolific causes, but to which morbid mental symptoms can, in fact, be rarely referred.   Insanity is never the pathological consequence of diseases of individual organs, but is occasionally more or less closely associated or connected with those forms of disease which result from diathesis or cachexia, such as tuberculosis, rheumatism, gout, and syphilis. There are many diseases painful in character and very depressing to the nervous system, such as calculus, fistula, cancer of the rectum and uterus, stricture with its often miserable complications, and many others which suggest themselves, which might be presupposed to be probable fertile causes of insanity, but which, in point of fact, are not specially inimical to brain health. They may be so indirectly, inasmuch as they prevent sleep, and produce anxiety and worry ; but even in this wise their effect is wonderfully slight. Nor does there appear to be sufficient evidence to warrant the connection of insanity with diseases of the heart, liver, or kidneys. It has been sought to show that certain forms of heart disease are associated occasionally with simple or hypochondriacal melancholy, and others with mania. These observations, however, are not supported by extended clinical observation. It is true that in certain rare cases, especially in connection with aneurism and aortic regurgitation, a maniacal delirium of short duration may present itself, due in all probability to anæmia of the brain ; but it is doubtful whether this condition can be classed as a true insanity, except for scientific purposes, on the principle that all morbid mental manifestations should be regarded as vesaniæ. Nor does it appear to me that diseases of the liver or kidneys have any direct connection with the induction of insanity, except perhaps that in some forms of Bright's disease a temporary mania is occasionally met with, probably the first indication of blood poisoning. The direct production of delirium, possibly passing the arbitrary line drawn between it and mania, has been often observed, but it is very doubtful whether prolonged mania or melancholia can be clearly shown to be associated with such diseases, except as producers of psychical over-exertion of the brain.

As regards cachexiæ and diatheses, we may at once accept the factorship of syphilis. Much more doubt surrounds the question of how far rheumatism and gout can induce true rheumatic or gouty insanity,—i.e., an insanity arising out of unphysiological conditions produced primarily in the nervous centres by the action of their toxic material. That they both produce alarming head symptoms, due probably to implication of the membranes of the encephale, we know ; but, so far as my experience goes, supported by reference to the literature of the two subjects, permanent mental aberration is unknown.   That they both may influence the course of a case of insanity may be freely admitted. The connective tissues of the brain and cord, predisposed by the diatheses to morbid change, no doubt now and then become implicated, probably in the immediate neighbourhood of blood-vessels, and by the consequent affection of sensory-motor areas and spinal segments, choreic movements are induced, and mental phenomena may result from variation of the local condition.   So far as I can ascertain, choreic movements present themselves *after* the incidence of insanity, which is hardly the sequence of events to be anticipated were the case truly one of rheumatic or gouty insanity.   The history of such cases leads to the conclusion that the poison in either case attacks regions undergoing degeneration or weakening by disease ; and, given vessels in a sub-inflammatory condition, the strong probability is that these toxic agents fasten on their connective tissues, and complicate the condition and the symptoms.   That we have not a true insanity resulting from these two diathetic conditions is all the more remarkable, as, in at least one of them, gout, mental phenomena in the form of irritability, depression, and paralysis of energy, are prodromata of the acute manifestations, disappearing for the most part with the appearance of local inflammation.   As our knowledge of the pathology of tuberculosis increases, stronger probabilities arise that it may be the direct cause of insanity ; but even in this case we have many secondary conditions to take into account, markedly anæmia, which prevent us coming to a definite conclusion on the subject.   There cannot be a doubt that the diathesis conditions the insanity of patients in which it is present, but this fact does not support the original proposition that the disease produces morbid mental symptoms.

This excursion is warranted in order to bring more prominently under view the fact that the *post hoc propter hoc* line of argument has to be most carefully guarded against in dealing with the diagnosis of the insanities.   It is of the utmost importance to the future of each case that its causation should be most carefully worked out; for not only is the original brain lesion different in kind according as it depends on over-exertion, traumatism, toxæmia, involutional and evolutional processes, and imperfect formation, but the line to be taken in treatment is indicated more or less distinctly by a knowledge of the efficient causes.

LECTURE V.

MR PRESIDENT AND GENTLEMEN,—Before taking up the question of treatment, you must allow me to state what appear to me to be the main conclusions arrived at in previous lectures. These are, first, that over-exertion of the areas of the brain which form the substrata of consciousness produces changes in anatomical and physiological relations evidenced by trains of physical and mental symptoms ; second, that the primary change is congestive in character ; and third, that secondary changes may be produced, resulting in impairment of the systems of connections and permanent insanity.

Were we dealing with the structure, function, and disease of any other organ, would not these conclusions be accepted as inevitable ? and, although differences of opinion as to details might arise, would not the relation of cause to effect be generally admitted ? Why is it that when the brain is approached a degree of hesitancy is manifested, expressive of reluctance to admit that perversions of its functions differ in some inscrutable manner from the results of pathological modifications of other systems ? Is it because the physiologist can offer no tangible explanation of what occurs during the interval between sensation and reaction—an interval which he (perhaps too willingly) resigns to the psychologist [1]—and can venture no hypothesis as to the nature of the processes accompanying transformations of specific energies ? If that be so, the same sort of glamour must surround all vital processes. The popular mind, however, accepts the didactic statements of the scientist on moot points, and attaches little importance to the hiatus which invariably exists in his lines of argument and experimentation ; not recognising the fact that as the processes going on in each system are investigated, the scientific observer finds himself at one period or

[1] Burdon Sanderson, " Presidential Address," British Association, 1893

another of his work on the edge of a gulf he cannot bridge. At
present, and it is to be feared for long, we must content ourselves
with the study of the conditions under which life can be maintained, in the hope that we may somewhat narrow the gap
between a knowledge of these conditions and a knowledge of
the nature of life itself.

To approach the treatment of the Insanities through the
portal of psychology is hopeless ; we have gained nothing by
taking that road in the past, and can hope for nothing in the
future. Much harm has been done by the prominence given to
a sort of pseudo-psychology by the specialty which professes the
treatment of the insane. It has kept up the idea contained in
the term " mental disease " by thought, word, and action. We
hear of medico-psychologists, psychiatric medicine, psychological
analysis, and so on. As a matter of fact, psychology has just as
much, and no more, to do with the treatment of insanity as it
has to do with the treatment of other forms of disease. We are
no more psychologists than you are ; and we know no more
about it than the physician who undertakes ordinary practice.
This affectation of a special knowledge of the science has done
the specialty much harm, and, through it, the insane. You have
to deal with perversions of the mental faculties day by day ; you
have to deal with hysterics, alcoholists, opium-eaters, so-called
neurasthenics, the weakness of old age, the phthisical, the febrile,
the pregnant, and endless other types in which modifications of
psychological functions are manifested. Do you apply to the
relief of such abnormalities the precepts of any special school of
psychology ? I think not. In every case of disease you meet
with, in addition to your special professional knowledge, you apply
the scheme of practical psychology which has grown up in each
and all of you, based on the experience of life. Does a man
know anything more about the workings of the component parts
of the mental constitution, and their attributes, because he has
practised more amongst the insane than the sane ? Has any great
scheme of psychology emanated from an asylum ? None that I
know of. In point of fact, the psychologist has invaded the
domain of physiology much more effectually than the so-called
psychiatric physician has that of psychology. I say this prominence given to psychology has done harm to the insane, for it

not only has kept in the background the study of the relations of the Institutes of Medicine to the study of the Insanities, but it has got apparently inextricably complicated with the system of treatment and management. Till Bevan Lewis published his important book on Insanity, no man in Great Britain ever even attempted to describe the anatomy of the organ the aberrations of which he was dealing with : many authors of such works prefaced their descriptions of symptoms by attempts to analyze the psyche, but till that time not one single author out of the dozens who have written on the subject had laid down the anatomy of a convolution. Under these circumstances it was inevitable that the study must have been conducted on psychological and clinical lines. I think it will be accepted as a postulate that the scientific knowledge of the nature of any disease depends primarily on a knowledge of the anatomy of the part affected. No one wishes to minimise the importance of clinical observation in the scientific scale, but in order that it may take its true position it must stand in direct relation to structure and function, otherwise it cannot rise above the level of empiricism. That our advances in this country have been mainly in the direction of clinicism is not the fault of men, but of systems, for the establishment of which they were in no way responsible.

You will perhaps remember that in the first lecture I speculated on the attitude the profession would assume to cases of Insanity presented to it for the first time ; and I asked you to try to imagine the position we would take up if, with our existing knowledge of brain anatomy and physiology and an extended acquaintance with other forms of disease, we were called on to diagnose and treat the patients so presented to us. I ventured the opinion that under these circumstances we would offer the provisional diagnosis of "disease of the brain with mental symptoms." We may now take a further imaginary step, and suppose that one or both of the cases died from some intercurrent disease, and that on post-mortem examination evidences of congestion, and its secondary consequences, were discovered. It would not be necessary in order to prove congestion to produce a cortex engorged with blood, in a so-to-speak erysipelatous condition ; were the products of inflammation found in the form of dilated vessels, extravasated leucocytes, pigment,

and marks of exudation, the evidence would be sufficient. I
submit that under these circumstances the natural history of the
disease would be complete, and that we would have before us a
pathological entity, whether we called it "congestive insanity,"
"acute mania," "acute melancholia," or "idiopathic insanity."
I think the last term is the best, as it expresses that the
symptoms had been produced by morbid processes commencing
primarily in the brain. It might be convenient to differentiate
further by speaking of "idiopathic mania," and "idiopathic
melancholia;" but if we fully understand what are the contents
of a term, it is of secondary importance what the term itself is.
The essence of any of these terms is over-stimulation, followed by
exhaustion of cell function, and production of certain mental and
physical abnormalities.

The principal deductions to be drawn from what has been
already said are twofold: first, that we must attack these
abnormalities through the organ primarily affected; and second,
that the first object to be obtained is REST for the brain. This can
be got by following the lines of treatment adopted in the case of
every-day disease. When we meet with a case in which we
have reason to believe that active implication of a special organ
will result not only in perversion of its own function, but will be
productive of degradation of other systems, we, as a rule, take
the preliminary precaution of confining the patient to bed, and
of keeping him there till such time as we procure reduction of
the primary morbid condition. In nine cases out of ten bed is
the best place to treat the prodromal stage of idiopathic
insanity. Further, should the case not have been submitted to
us till excitement of a maniacal or melancholic type has
developed itself, or our treatment of the early stage not have
been entirely successful, bed is still the proper place to conduct
treatment. In the incipient case the person will be very ill
if the disease is not arrested; in the acute case the person *is*
very ill. Why this emphasis? Because a very usual practice is
to prescribe entirely different conditions for treatment. Acting
on the psychological principle that the mind is affected and
requires to be diverted, a common custom is to order foreign
travel, a long voyage, change of scene in degree and kind
according as the circumstances of the patient permit. In the

very large proportion of cases these procedures appear to me (to put it roughly) as inadmissible as it would be to order a patient suffering from incipient pneumonia to walk up a mountain side. Let us look at the position from both aspects—the psychological and the pathological. We know that the exciting cause is worry, *i.e.*, a constant succession of painful emotions, which has produced the condition of the brain already indicated. But it may be asked, how are we to discriminate between this state and an attack of low spirits so constantly observed? There is a very distinct line of demarcation between depression of feeling within the limits of health, and the depression of feeling resulting from morbid processes going on in the brain. In the one we have a function exercised to its uttermost, yet never diseased; in the other we have a function exercised beyond normality, beyond the power of the organ to obtain immediate recuperation. But between the two there is a distinct line of demarcation. A mere fit of "low spirits," from whatever cause, does not prevent a man from using his intellectual faculties. Circumstances influence him; he can review his position, notwithstanding that he does so gloomily, and the turn of events and the moderating influence of time produce restoration of physiological and psychological equilibrium. In this case the brain has evidently not been diseased; excitation has not produced morbid processes. The symptom which indicates that the brain is distinctly affected, that the limit of health has been passed, is modification of the natural influence of external circumstances, conveying to the pathologist indications of implication of the integrity of the brain elements. This is further expressed by the falling off in bodily condition. We all know the general condition better than I can express it in words; why then should we be so prone to merge the two in one when we come to devise measures for their relief? It is true, rest is demanded in either case. *Mais il y a fagots et fagots.* In the one the rest which comes from a change in *the character* of stimulus is called for; in the other the rest which comes from *cessation* of stimulus, so far as it is possible to obtain it, is demanded. The two conditions were probably represented by the second and fourth conditions of Hodge's cells,—the one being in a state admitting of rapid repair, the other having so far lost its pabulum as to prevent the normal exercise of function, and to require a long period for its reconstitution. Stimulation is not to be reduced by subjecting the patient to the excitement and worry of travelling, taking travel as an instance of psychological methods of treatment ; nor is the tendency to degradation of the general system likely to be arrested. It is not a scheme of treatment which would be adopted in analogous conditions of any other organ. I am far from saying that in certain forms of insanity such a system of treatment is not advisable ; in certain degenerative cases it may be employed, but even in these the utmost caution is necessary ; but in all cases in which there is reason to believe that active disease of the convolutions is imminent, or actually in progress, rest of the most perfectly attainable character is urgently called for, and this rest is best obtained in bed. I may be asked, does not the patient resent such confinement? My answer is, rarely in my experience. You may remember that in a former lecture I said the patient in the initial stage is often painfully aware of the nature of his malady—that he dreads the incidence of insanity. He is conscious that he is unwell, that he does not react normally to external circumstances, that he cannot apply himself to the everyday business of life. The explanation of this fact must be left to the psychologist, but that it is a fact must be supported by the experience of many present. Under these circumstances the patient finds comfort in accepting the position, and in the assurance that if he will submit himself to treatment for a few weeks, the brain condition which is the cause of his malady will be overcome. I quite see the apparent contradiction of the statements that patients in the condition we are speaking of do not react normally to external circumstances, and that they can accept a position. I can only take refuge in the clinical fact that it is so. It may be in either case a matter of degree ; abnormalities have many degrees of intensity, and it may be that refusal and acceptance are both imperfect. There is no doubt that the sane has a powerful effect on the insane mind, except in extreme cases. In considering the treatment and management of the insane this must be always kept in view. It is perversion, not destruction, of function we have to deal with, consequent on changes in, but not destruction of, the organ. It is a matter of *derangement* of both organ and function. Although deranged they may work imperfectly, or one part may act and another not.   Still a certain power of work is evidently left.   Few men are perfectly insane ; many insane persons can listen to argument, and be more or less influenced by it.   Have we not seen the victim of *delirium tremens* pull himself together for a few minutes ?   In an incipient case of idiopathic insanity mental power is not lost, and the patient is, in nine cases out of ten, able to submit himself to discipline.   The influence of a stronger mind always tells on a pathologically weaker one.   I need not tell you that the physician is most powerful in the sick-room or hospital, and that he strengthens his position when he orders the patient to bed.   Even in cases which at first resent the order, as soon as it is obeyed no difficulty arises in maintaining it.   A good nurse, or even two may be necessary.   The ordinarily trained nurse is sufficient, unless considerable excitement intervenes, in which case the so-called "mental" nurse is desirable ; but she must have had hospital training.   I do not intend here to go into the question of the nursing of the insane, further than to say that all "mental" nurses should have been trained in a general hospital, or under a hospital nurse in an asylum.   Good as the old "attendant" may have been, she never appreciated the fact that her "mental" patient was an invalid first and an insane person after.   This the hospital trained attendant appears fully to understand ; her patient is to her a sick man or woman,—the insanity is an incident in the case.   This is due to her training ; which has also engendered in her a stricter sense of duty, a knowledge of various important manipulations, and increased power of observation.   Such nurses can and do tend men unless excitement supervenes.   Would that some scheme could be instituted by which male nurses could be instructed in the same manner as women ; were it only because one man-nurse could often manage, where it is for obvious reasons necessary to employ two women.   As a rule, however, we are compelled to call in the male nurse as we find him, and if he has been trained as he should have been, he will do well enough.   With such a staff the physician has his patient practically in his power, even in the case of one who enters protests against the measures taken.   The protests soon die away, the physician's position is established, and the patient is under circumstances proper for the application of therapeutic agents. One word before going further,—as to surroundings: Is it necessary to remove the early case from home to lodgings in the country? As a rule, I think not. If he is confined to his own room, and kept entirely away from his family or friends, it is enough. He should be carefully secluded, and it may be even desirable that the rest of the family should leave the house. Visits stimulate; further, they encourage discussions on the propriety of the physician's line of management, and so reduce discipline.

The object to be obtained is sleep, for it is the first indication of returning brain health. We are dealing with congestion, and the return of sleep is the physiological sign that the cerebral vessels are re-acquiring their power of contractility. The observations of Durham and Jackson, taken along with those of Mosso and others, show that during sleep the brain vessels of the superior convolutions shrink. I must here enter a protest against the common practice—one possibly forced on the physician by the pressure of circumstances—of at once resorting to the administration of hypnotic drugs. We have seen the effects of prolonged stimulation of cells, and to what an extent of exhaustion they may be reduced. Let me direct your attention to certain experiments by Binz of Bonn. Taking the brains of recently killed cats and rabbits—the organs, in fact, being removed immediately after death—he placed small portions of the convolutions in (1), 0·7 per cent. solution of sodium chloride, (2), 0·2 per cent. solution of atropine sulphate, and (3), 0·2 per cent. solution of morphine sulphate, in which they were allowed to soak for fifteen minutes. The specimens were then squeezed out under cover glasses, and examined with the microscope. Those submitted to the sodium chloride and atropine sulphate solutions showed the cells scarcely, if at all, altered; whereas the cells of the portion steeped in the morphine solution had become affected, the body presenting an opalescent appearance suggestive of the commencement of evaporation. Dilute acids, lactic and nitric, 1 to 3000, had the same effect. Very dilute solutions of morphine, 1 to 5000, produced the same appearances, but not so distinctly. Chloral hydrate, chloroform, and ether gave somewhat similar results; but atropine, caffeine, camphor, and some other substances appear to have had no effect. He further found by examining the exposed brains of animals which had been narcotized by chloral hydrate, ether, and chloroform, that there was no immediate change in the vascularity of the organ, but that after full narcosis had been obtained it became anæmic.[1] Binz's observations led him to the conclusion that narcotics acted directly on the cells, and that narcosis was not produced by any constricting influence on the bloodvessels. Although these experiments cannot be held as conclusive, they are still highly suggestive, and it seems as if this line of inquiry might be followed up with advantage. What the appearance of commencing coagulation is due to is not evident ; but if morphia compounds and chloral hydrate have any direct action on healthy cells, all the more may they be supposed to act detrimentally on cells degraded in constitution and function. This seems borne out by clinical observation. Although sleep can be obtained by their administration,—generally in larger doses than suffice for the sane subject,—on awakening the condition of the patient is not improved, to say the least. My old master and friend Skae of Morningside used to lay it down as an axiom, that opium in any shape might reduce the immediate intensity of symptoms, but it prolonged their duration. If we depend on Binz's observations to any extent, we obtain some indication of a reason for this opinion,—the poisoning of an already reduced organ. We have further to take into account the effects most narcotics, especially opium, have on the alimentary apparatus, which are, as has been stated, generally out of tone in the class of patients we are dealing with ; the liver secretions are affected, and the general atony is increased, shown by a foully furred tongue, offensive breath, and enormous evacuations For every reason opiates are contra-indicated ; chloral hydrate is not so harmful, still it and all other narcotics should be avoided. Sleep may be attained by much simpler methods.

The method of action of counter-irritation is still very much a matter of surmise, and it is not for me to endeavour to elucidate the subject. Long clinical experience tells us that in many inflammatory and sub-inflammatory conditions its application

---

[1] *Archiv für Experimentelle Pathologie,* Band vi.

beneficial. I have found it so in the class of cases we are now considering. It is my custom to apply mustard on the upper part of the chest and of the back for long-continued periods, placing such a poultice on the front one night and on the back the next, alternating the application so long as the skin will bear it without vesication. It is a matter of indifference what rubefacient is employed, but mustard leaves are the most convenient. Uneasy feelings in the head are soon relieved by this process, and it not unfrequently happens that sleep for shorter or longer periods follows. Blistering is an unnecessarily severe measure. Warmth maintained by bottles applied to the feet and back, and by warm compresses over the abdomen, is often useful. Very gentle shampooing or *massage* applied to the head and neck four or five times a day for a few minutes at a time is generally soothing. This should be done by the spread-out fingers, starting from the sides and back of the head, and being carried gently down to the clavicles and the root of the neck all round. Pressure should be slight. It is possible that this process may relieve the veins and lymphatics of the neck, and consequently those of the brain. Food should be given in small quantities, and often,—every two or three hours, the largest meal being at night. As I am speaking now of food only as a sleep-producer, the general question of diet may lie over for a time. The best method of bleeding a man's brain is to draw the blood into the vessels of the abdomen by means of food ; and therefore frequency of meals is called for, and the night meal should be the most plentiful. Of drugs, digitalis is the most useful, given in fairly large doses, and pretty frequently,— $\mathfrak{M}$ 15 doses every two or three hours. In this disease there is considerable toleration of this and other drugs. Digitalis has no direct action on the nervous system ; in cases of poisoning the intellect is in no way affected. Its action on the cells is through the vascular system, by producing contraction of the terminal vessels. When sleep follows rapidly on its administration in conjunction with the other measures just alluded to, the prospect is very hopeful, for it indicates that convolutional congestion is being relieved. That this drug acts on the terminal vessels is opposed to the views of Schmiedeberg ; it is, however, in accordance with the teaching of Lauder Brunton, and my clinical observation certainly goes to support the views of the latter.

Sheridan Lea, in his *Chemical Basis of the Animal Body*, describes neurin as a product of the decomposition of protogon, a substance obtained from the brain by Liebreich, and regarded by him as its principal constituent. Brieger has shown that " neurin is one of the most commonly occurring and actively toxic alkaloidal basic products of the putrefactive decomposition of animal tissues known under the name of ptomaines." Noël Paton tells me that on trying to draw up a statement of the decomposition products of nerve tissue, he finds the existing knowledge of the processes and of the products is so indefinite that nothing of positive value can be given ; all he can say is that our present knowledge suggests how poisonous products may be developed. Under these circumstances, it is somewhat drawing a bow at a venture to suggest that neurin is the poison of exhausted cells productive of the symptoms of fatigue shown in Mosso's experiment cited in the second lecture, and of exudates due to congestion. Still I offer the suggestion. It may, therefore, be desirable to maintain diuresis, as all such compounds are removed by the kidneys. This may be one of the benefits accruing from the administration of digitalis.

Steady perseverance in this system for a week, or even less, will generally produce natural sleep. As a rule, the patient begins to sleep by snatches, and, as the case goes on, in more continuous periods. No doubt steady sleep for five or six hours, or even longer, is what we hope for ; but if a patient, who for days or weeks previously has hardly slept at all, and then only probably by the use of alcohol or opiates, gets five or six hours in the twenty-four by an hour or two hours at a time, he does fairly well. And between such naps there is often a calm apathy, an indifference, and lassitude, which are most useful periods of rest. Perhaps it is a mild obfuscation corresponding to the period of dementia which is so frequently observed as a *sequela* of acute attacks of mania and melancholia, a state of brain weakness, still the weakness of convalescence.

It is possible, however, that although the symptoms may become modified, they may not disappear, and the bodily con-

dition may not materially improve. If the temperature shows
even a slight evening rise, it is an indication that mischief is
still going on.   For the last three years I have, under these circumstances, used phenazonum (antipyrin) freely.   In doses of
15 grains, given every two hours, it very generally soothes and
produces sleep, and the general condition improves.   We have
no evidence before us as to the manner in which this drug acts
on nerve cells : its influence in checking metabolism, as shown
by its power of arresting the secretion of urea,[1] and its vaso-
constrictor action in small doses, may be the efficient causes.
Although I have frequently given antipyrin to the amount of
60 to 100 grains per diem for a fortnight at a time, I have never
noticed any of the ill effects ascribed to it, except occasionally
nausea and vomiting.   It may be that patients in the condition
we are speaking of may be more tolerant of the poison than
those in whom temperature runs high.   As in all nervous cases,
the dosage is at first a matter of experiment ; but it is quite safe
to begin with the dose above indicated, raising it somewhat, and
administering it more frequently according as the patient's condition improves.

There is yet another means of bleeding the encephale,
which has also an important action on the general systemic
nutrition of persons suffering from cortical cell weakness.
I allude to the functional hyperæmia of muscular activity.
Muscular exercise excites the circulation generally, and this
should be avoided.   Restlessness, under ordinary circumstances,
cannot be overcome by an amount of exercise short of fatigue.
The sufferer from the incipient symptoms of idiopathic insanity
is not in a condition to stand fatigue ; his nervous system is
undergoing waste, and to make further demands on it is pathologically and physiologically illogical.   It is not only undergoing
waste itself, but it is, as a consequence of its own wasting condition, unable to exercise its nourishing functions over the system
at large.   Fatigue in any sense or in any degree is to be depre-
cated.   But with the patient in bed we would lose not only the
important derivative action of hyperæmia incident on muscular
action, but the products of muscular waste which, to a certain
extent, must accumulate under restful circumstances, would

[1] Stockman, *New Official Remedies*, 1890.

remain in the tissues, except so far as they may be removed by the almost passive action of veins and lymphatics. This hyperæmia can, however, be easily attained without making any serious call on the nervous system. Gentle shampooing of the whole surface of the body with the open hand will suffice. We require none of the refinements of the school of *massage;* only light hands, some judgment, and obedience to orders on the part of the nurses. The open hand should work gently upwards from the toes and fingers to the hip and shoulder, the abdomen should be kneaded ; if this is well done—that is to say, if it does not excite the patient, as it sometimes does even those who are not nervously affected—the open hand may be succeeded by gentle kneading of the limbs. It usually is advisable to be very cautious in the employment of the shampoo, beginning with not more than five minutes at a time, and gradually increasing it to a quarter of an hour three or four times a day, according as the patient stands it. I find it better not to use it after six o'clock at night over the body, after which hour the head and neck massing only should be exercised. It is less apt to excite if oil is used. When there is a tendency to emaciation, oil gently rubbed into the skin often materially assists nutrition. Under such circumstances I not only cause the oil to be rubbed in, but the patient is advised to sleep without sheets, between blankets or in a flannel night-gown, which become more or less saturated with the oil. This sounds unpleasant, but those of us who have adopted similar measures in tubercular patients know that it is not resented. A warm bath in the evening, at a temperature not higher than 97°, materially promotes the tendency to sleep. Mild catharsis should be maintained.

If the case fails to react to these measures—it rarely does so —if sleep is not induced, its aspect is grave ; and it may be necessary to administer an hypnotic. This, however, should be delayed as long as possible, and should be regarded as an extreme measure. Cases are often sent to asylums whose treatment at home has been mainly conducted by the administration of opiates and other soporifics, and which are treated to a successful termination without the exhibition of any such drugs. Force of family circumstances has compelled their administration. The least harmful soporific in my experience is paraldehyde, administered in 2 to 3 drachm doses per rectum. In this manner the drawback of its unpleasant taste is overcome. Sulphonal has been much recommended, but in exhausted cases its paretic effects, which may be shown only by a slight tired feeling of the legs, renders it, in my opinion, an inferior sleep-producer to paraldehyde. I have no faith in organic extracts, or in hypnotism. The employment of the latter has, in two instances within my own experience, resulted in the production of grave evils.

Lastly, as to diet. A plain everyday medium sick-room diet, given in small quantities and often, is all that is demanded. Good soups, fish and fowl, eggs, milk, and vegetables, in judicious admixture, suffice. The poison of trade meat extracts should be avoided. I never could see the object of stuffing a patient whose digestive system is lowered in function with an amount of egg, oil, and cream that would be resented by the stomach if taken under circumstances of full health. Such a course might be advisable for a troop of soldiers about to set out on an arduous campaign, in whom reserve force should be stored up ; but in our case reserve is not called for, as we can supply nourishment hour by hour in quantities which can be easily assimilated. We have to build up gradually, working on organs which are not fitted for extra exertion. General fattening of the tissues is no doubt a sign of returning health, but the mere local accumulation of fat is not. Fat follows on health, not health on fat.

As soon as we have distinct evidence of improvement in returning sleep, bodily betterment, and mental composure, although the last-named may be accompanied by stupidity, nervine tonics are of value. Till then they are useless. Without any belief in the axiom, *Ohne Phosphor keine Gedanke*, the phosphates have succeeded best in my hands. Taken over all, the Easton's syrup of the phosphates, omitting the quinine, has worked best. In certain instances 5-minim doses of liquor arsenicalis, substituted for the quinine, has been highly beneficial. For reasons stated in last lecture, emmenagogues are useless ; the function returns as the balance of health of the system as a whole is becoming re-established. In like manner the direct treatment of oxaluria and phosphaturia is of no value. Along with returning health comes the time for change of scene, travel, and other means of occupation or amusement. Till then they are worse than useless; they form another class of the tonics of convalescence.

Need it be said that any general line of treatment must, as in the case of every type of disease, be modified according to cases and circumstances? It may be said that I have spoken only of the treatment of the patient who can afford it, and that it is impossible to carry out such a system in the case of the poor man. It must be admitted that in a large number of such instances little or nothing can be done, but in many the goodness of the poor to each other helps us to treat such patients in a more or less satisfactory manner. Were it not that our general hospitals are closed by special regulations against any cases presenting mental symptoms unless they are produced by alcohol, many a case of incipient insanity would be cured, relegation to an asylum would be unnecessary, and the accumulation of pauper lunatics might to no small extent be stayed. I have no hesitation in saying that were wards set aside in each of our general hospitals for the treatment of early cases of Insanity by the ordinary physicians, a vast amount of misery would be relieved, and, what appeals to a large section of the public, the yearly increasing expenditure on asylums would in no slight measure be obviated.

But let the sufferer from incipient Idiopathic Insanity be rich or poor, he must be treated on hospital principles, whether in his own house or in an institution; and he must submit to be placed in the hands of his physicians and nurses absolutely. The treatment may be irksome, but it is not more so than that necessary for the relief of many other morbid conditions. From three to six weeks is generally the time necessary to procure convalescence, and another period of from one to three months may be needed for complete recovery, i.e., before a person is fit to return to regular work. Even then business should be commenced with caution.

Mr President and Gentlemen, I know what I have said as to treatment is commonplace, and I hope you have regarded it as such. I could easily have submitted to you clinical cases illustrative of symptoms, treatment, and results; but it would only have been to ring the changes on the statements made *en masse*. I have endeavoured to give a commonplace description of a commonplace type of disease up to a certain point, bearing on points of its treatment according to their prominence in my own estimation. If I have advanced one argument, or produced one item of evidence, which has enabled any of you to reduce the condition to the ordinary level of everyday disease, to divest it of any mystery hanging about it, or to see that its treatment is not characterized by any inherent difficulty, my object will have been gained. I submit that the view taken of the disease by the profession under the imaginary circumstances which we pictured, would have been of the nature I have endeavoured to lay before you. Although the mental symptoms are to the jurist and the citizen the essence of the case, to the physician they are of secondary importance. They are mere incidents dependent on certain conditions of the superior convolutions. Let me adduce a case in point, that of a young man now under treatment whose symptoms ran the following course : gradual but great falling off in bodily health, during which time he was peculiar, restless, irritable, and unable to work ; delusion of a religious character supervened ; ecstasy; perversion of natural affection ; apathetic melancholia ; acute mania ; and convalescence. This is not a common sequence of symptoms, but its uncommonness strengthens the argument. We have every right to refer the various symptoms to a common cause, and to conclude that they were incidents springing from incidents in the course of the morbid cerebral process. I know my statements as to this point have been much misrepresented. That is a small matter. Symptoms have their own special value, and that a very high one ; but they are not the disease. Until the student of so-called "mental disease" estimates their due relation to anatomy, physiology, and pathology, until he determines their proper value in the equations he is daily striving to solve, so will the terms mania, melancholia, and dementia remain undetermined factors, and insanity an empirical expression.

It was my intention to discuss the treatment of active idiopathic mania and melancholia during this course of lectures ;

but the various demonstrations have occupied so much time that their consideration must be deferred to some other occasion. I submit, however, I have fulfilled the promise made in the first lecture to discuss the treatment of the early conditions produced by over-exertion, starting from an anatomical, physiological, and pathological basis.

PRINTED BY OLIVER AND BOYD, EDINBURGH.